Relativistic Quantum Information

Relativistic Quantum Information

Ignazio Licata
60 Festschrift

Editors
Fabrizio Tamburini
Ignazio Licata

MDPI • Basel • Beijing • Wuhan • Barcelona • Belgrade • Manchester • Tokyo • Cluj • Tianjin

Editors
Fabrizio Tamburini
Zentrum für Kunst und Medientechnologie
Germany

Ignazio Licata
ISEM Institute for Scientific Methodology
Italy

Editorial Office
MDPI
St. Alban-Anlage 66
4052 Basel, Switzerland

This is a reprint of articles from the Special Issue published online in the open access journal *Entropy* (ISSN 1099-4300) (available at: https://www.mdpi.com/journal/entropy/special_issues/relativistic_quantum_information).

For citation purposes, cite each article independently as indicated on the article page online and as indicated below:

LastName, A.A.; LastName, B.B.; LastName, C.C. Article Title. *Journal Name* **Year**, *Article Number*, Page Range.

ISBN 978-3-03943-260-8 (Hbk)
ISBN 978-3-03943-261-5 (PDF)

© 2020 by the authors. Articles in this book are Open Access and distributed under the Creative Commons Attribution (CC BY) license, which allows users to download, copy and build upon published articles, as long as the author and publisher are properly credited, which ensures maximum dissemination and a wider impact of our publications.

The book as a whole is distributed by MDPI under the terms and conditions of the Creative Commons license CC BY-NC-ND.

Contents

About the Editors vii

Preface to "Relativistic Quantum Information" ix

Ignazio Licata
Some Notes on Quantum Information in Spacetime
Reprinted from: *Entropy* **2020**, *22*, 864, doi:10.3390/e22080864 1

Lawrence Crowell and Christian Corda
Quantum Hair on Colliding Black Holes
Reprinted from: *Entropy* **2020**, *22*, 301, doi:10.3390/e22030301 5

Fabrizio Tamburini and Ignazio Licata
General Relativistic Wormhole Connections from Planck-Scales and the ER = EPR Conjecture
Reprinted from: *Entropy* **2020**, *22*, 3, doi:10.3390/e22010003 21

Stefano Liberati, Giovanni Tricella and Andrea Trombettoni
The Information Loss Problem: An Analogue Gravity Perspective
Reprinted from: *Entropy* **2019**, *21*, 940, doi:10.3390/e21100940 35

Ben Maybee, Daniel Hodgson, Almut Beige and Robert Purdy
A Physically-Motivated Quantisation of the Electromagnetic Field on Curved Spacetimes
Reprinted from: *Entropy* **2019**, *21*, 844, doi:10.3390/e21090844 65

Adrian Kent
Summoning, No-Signalling and Relativistic Bit Commitment
Reprinted from: *Entropy* **2019**, *21*, 534, doi:10.3390/e21050534 91

Xiaodong Wu, Yijun Wang, Qin Liao, Hai Zhong and Ying Guo
Simultaneous Classical Communication and Quantum Key Distribution Based on Plug-and-Play Configuration with an Optical Amplifier
Reprinted from: *Entropy* **2019**, *21*, 333, doi:10.3390/e21040333 101

Shujuan Liu and Hongwei Xiong
On the Thermodynamic Origin of Gravitational Force by Applying Spacetime Entanglement Entropy and the Unruh Effect
Reprinted from: *Entropy* **2019**, *21*, 296, doi:10.3390/e21030296 115

About the Editors

Fabrizio Tamburini, Ph.D, works on Electromagnetic Orbital Angular Momentum (OAM). His scientific and artistic contributions include OAM telecommunications, superresolution, OAM vorticities from rotating black holes and axion dark matter.

Ignazio Licata, born 1958, is an Italian theoretical physicist, Scientific Director of the Institute for Scientific Methodology, Palermo, and Professor at School of Advanced International Studies on Theoretical and Nonlinear Methodologies of Physics, Bari, Italy, and Researcher in International Institute for Applicable Mathematics and Information Sciences (IIAMIS), B.M. Birla Science Centre, Adarsh Nagar, Hyderabad 500, India. His research covers quantum field theory, interpretation of quantum mechanics and, more recently, quantum cosmology. His further topics of research include the foundation of quantum mechanics, dissipative QFT, space-time at Planck scale, the group approach in quantum cosmology, systems theory, nonlinear dynamics, as well as computation in physical systems (sub- and super-Turing systems). Licata has recently developed a new approach to quantum cosmology with L. Chiatti ("Archaic Universe") based on de Sitter group, and has proposed a new nonlocal correlation distance, the bell length, with D. Fiscaletti.

Preface to "Relativistic Quantum Information"

Relativistic quantum information (RQI) is a multidisciplinary research field that involves concepts and techniques from quantum information with special and general relativity. General relativity and quantum physics are two established domains of physics that have been mutually incompatible until now. Hawking radiation, the black hole information paradox including soft photons and gravitons, the equivalence between the Einstein–Rosen bridge from general relativity, and the Einstein–Podolski–Rosen paradox from quantum mechanics are examples of the new phenomena that arise when two theories are combined. RQI uses information as a tool to investigate spacetime structure. On the other hand, RQI helps to identify the applicability of quantum information techniques when relativistic effects become important: entanglement and quantum teleportation can be used to reveal gravitational waves or realize a quantum link between satellites in different reference frames in view of future large-scale quantum technologies. The aim of this Special Issue is to take stock of state-of-the-art perspectives on RQI, with particular attention to the concept of quantum information and the repercussions of RQI on the foundations of physics.

Fabrizio Tamburini, Ignazio Licata
Editors

Editorial

Some Notes on Quantum Information in Spacetime

Ignazio Licata [1,2]

1 ISEM, Institute for Scientific Methodology, 90121 Palermo, Italy; ignazio.licata3@gmail.com
2 School of Advanced International Studies on Applied Theoretical and Non-LinearMethodologies in Physics, 70121 Bari, Italy

Received: 15 July 2020; Accepted: 30 July 2020; Published: 6 August 2020

The results obtained since the 70s with the study of Hawking radiation and the Unruh effect have highlighted a new domain of authority of relativistic principles. Entanglement, the quantum phenomenon par excellence, is in fact observer dependent [1], and the very concept of "particle" does not have the same information content for different observers [2,3]. All this proposes the centrality of the notion of "event" in physics and the meaning of its informational value. It is in this direction that Quantum Relativistic Information (QRI) is defined, which can therefore be defined as the study of quantum states in a relational context.

It must be said that, despite being a prelude to a future quantum gravity, QRI is a largely autonomous field—because it does not imply any specific hypothesis on the Planck scale—and is characterized by some principles that guard an assumption of great epistemological strength. As A. Zeilinger [4] says, it is impossible to distinguish between "reality" and "description of reality", i.e., information in the study of physics; doing so means jeopardizing the universal value and beauty of physical laws. Both relativity and quantum physics are aspects of a broader information theory that we have been discovering in recent years and within which the foundational debate is renewed with new experimental possibilities. The first principle we need is therefore:

The principle of contextuality [5]: Each description of a class of events must contain, implicitly or explicitly, the reference structure of the observer. In other words, it must be possible for each observer to define assign values for each observable.

A very strong request comes from the principle of equivalence, which, after showing unsuspected resistance to any attempt of de-construction, is now extended to the quantum domain as a request to describe gravitational phenomena in terms of causal networks [6–11]. L. Susskind and G. 't Hooft proposal for the information paradox adds a new element to the picture: the complementarity invoked is in fact a principle of equivalence [12,13]. Although the Black Holes question are still far from being resolved (with particular regard to the core of the BH, with interesting inter-connections between strings, non-commutativity and euclidicity, see for example: [14–20]), the synthesis of equivalence and complementarity leads to a powerful holographic principle that introduces, according to Bekenstein's limit [21], a new way of looking at the locality and a different approach to cosmology. The holographic principle feeds on conjectures and is still looking for theories (duality between gravity and quantum field theory: [22–26]), but it is a catalyst for new conceptual suggestions regarding the physical meaning of the cosmological horizon. In particular, considering the four-dimensional dynamics as the explication (in a Bohmian sense) of a De Sitter non-perturbative vacuum offers an improvement of Hartle–Hawking proposal in quantum cosmology and a solution to the informational paradox in the BH [27–29]. This line of reasoning is also promising for an event-based reading of Quantum Mechanics [30].

For a long time, holography and emergentism appeared as two styles of explanation irreconcilable with respect to the locality, but an emergency of time could offer new perspectives with a duality between imaginary time and real time, in a diachronic/synchronic complementarity [31–33].

It is known that there are well-defined whormhole solutions in General Relativity and Yang Mills Theory, and the recent ER = EPR conjecture proposes the question of the emergence of metric space-time from a non-local background [34–38]. A suggestion in the direction of the laboratory comes from the Bose–Marletto–Vedral conjecture on the possible coalescence of two quantum systems in a non-local phase, which would reveal the limits of the local metric description and the non-classical aspects of space-time [39,40]. A covariant analysis of this situation shows that discrete effects could prove to be an overlap of geometries measurable through entanglement entropy [41,42].

Furthermore, localization appears as the production of a new degree of freedom. We assume, in accordance with a recent proposal [30,43], that the localization R of a process is associated with the genesis of a micro-horizon of de Sitter of center O and radius $c\theta_0 \approx 10^{-13}$ cm (chronon, corresponding to the classical radius of the electron), with O generally delocalized according to the wave function entering/leaving the process. The constant θ_0 is independent of cosmic time, so the ratio $t_0/\theta_0 \approx 10^{41}$ is also independent of cosmic time, with $ct_0 \approx 10^{28}$ cm. This ratio expresses the number of totally distinct temporal locations accessible by the R process within the horizon of cosmological de Sitter. In practice, the time line segment on which an observer at the center of the horizon places the process R has length t_0, while the duration of the process R is in the order of θ_0; the segment is therefore divided into separate $t_0/\theta_0 \approx 10^{41}$ "cells". Each cell can be in two states: "on" or "off". The temporal localization of a single process R corresponds to the situation in which all the cells are switched off minus one. Configurations with multiple cells on will correspond to the location of multiple distinct R processes on the same time line. If you accept the idea that each cell is independent, you have $2^{10^{41}}$ distinct configurations in all. The positional information associated with the location of 0, 1, 2, ... , 10^{41} R processes then amounts to 10^{41} bits, the binary logarithm of the number of configurations. This is a kind of coded information on the time axis contained within the observer's de Sitter horizon.

The R processes are in fact real interactions between real particles, during which an amount of action is exchanged in the order of the Planck quantum h. Therefore, in terms of phase space, the manifestation of one of these processes is equivalent to the ignition of an elementary cell of volume h^3. The number of "switched on" cells in the phase space of a given macroscopic physical system is an estimator of the volume it occupies in this space, and therefore of its entropy. It is therefore conceivable that the location information of the R processes is connected to entropy through the uncertainty principle. This possibility presupposes the "objective" nature of the R processes.

It is therefore natural to ask whether some form of Bekenstein's limit on entropy applies in some way to the two horizons mentioned. If we assume that the information on the temporal location of the processes R, $I = 10^{41}$ bits, is connected to the area of the micro-horizon, $A = (c\theta_0)^2 \approx 10^{-26}$ cm^2 from the holographic relationship:

$$\frac{A}{4l^2} = I \qquad (1)$$

Then, the spatial extension l of the "cells" associated with an information bit is $\approx 10^{-33}$ cm, the Planck scale! It is necessary to underline that the Planck scale presents itself in this way as a consequence of the holographic conjecture (1), combined with the "two horizons" hypothesis, and therefore of the finiteness of the information I. It in no way represents a limit to the continuity of spacetime, nor to the spatial or temporal distance between two events (which remains a continuous variable). Furthermore, since $I = t_0/\theta_0$ and t_0 is related to the cosmological constant λ by the relation $\lambda = 4/3t_0^2$, the (1) is essentially a definition of the Planck scale as a function of the cosmological constant. A global-local relationship is exactly what we expect from a holographic vacuum theory.

Funding: This research received no external funding.

Conflicts of Interest: The author declare no conflict of interest.

References

1. Alsing, P.M.; Fuentes, I. Observer dependent entanglement, Class. *Quantum Grav.* **2012**, *29*, 224001. [CrossRef]
2. Davies, P.C.W. Particles do not exist. In *Quantum Theory of Gravity*; Essays in Honor of the 60th Birthday of Bryce DeWitt; Christensen, S.M., Ed.; Adam Hilger: Bristol, UK, 1984; pp. 66–77.
3. Colosi, D.; Rovelli, C. What is a particle? *Class. Quant. Grav.* **2009**, *26*, 025002. [CrossRef]
4. Zeilinger, A. *Dance of the Photons: From Einstein to Quantum Teleportation*; Farrar Straus Giroux Publisher: New York, NY, USA, 2010.
5. Jaroszkiewicz, G. Observers and Reality, in Beyond Peaceful Coexistence. In *The Emergence of Space, Time and Quantum*; Licata, I., Ed.; Imperial College Press: London, UK, 2016; pp. 137–151.
6. Licata, I.; Corda, C.; Benedetto, E. A machian request for the equivalence principle in extended gravity and non-geodesic motion. *Grav. Cosmol.* **2016**, *22*, 48–53. [CrossRef]
7. Licata, I.; Benedetto, E. The Charge in a Lift. A Covariance Problem. *Gravit. Cosmol.* **2018**, *24*, 173–177. [CrossRef]
8. Candelas, P.; Sciama, D.W. Is there a quantum equivalence principle? *Phys. Rev. D* **1983**, *27*, 1715. [CrossRef]
9. Tamburini, F.; de Laurentis, M.F.; Licata, I. Radiation from charged particles due to explicit symmetry breaking in a gravitational field. *Int. J. Geom. Methods Mod. Phys.* **2018**, *15*, 1850122. [CrossRef]
10. Zych, M.; Brukner, C. Quantum formulation of the Einstein equivalence principle. *Nat. Phys.* **2018**, *14*, 1027–1031. [CrossRef]
11. Hardy, L. Implementation of the Quantum Equivalence Principle. *arXiv* **2019**, arXiv:1903.01289.
12. Hooft, G. Dimensional Reduction in Quantum Gravity. *arXiv* **1993**, arXiv:gr-qc/9310026.
13. Susskind, L. The paradox of quantum black holes. *Nat. Phys.* **2006**, *2*, 665–677. [CrossRef]
14. Susskind, L. String Physics and Black Holes. *Nucl. Phys. Proc. Suppl.* **1995**, *45*, 115–134. [CrossRef]
15. Strominger, A.; Vafa, C. Microscopic Origin of the Bekenstein-Hawking Entropy. *Phys. Lett.* **1996**, *379*, 99–104. [CrossRef]
16. Yogendran, K.P. Horizon strings and interior states of a black hole. *Phys. Lett.* **2015**, *750*, 278–281. [CrossRef]
17. Nicolini, P. Noncommutative Black Holes. The Final Appeal to Quantum Gravity: A Review. *Int. J. Mod. Phys.* **2009**, *24*, 1229–1308. [CrossRef]
18. Dowker, F.; Gregory, R.; Traschen, J. Euclidean Black Hole Vortices. *Phys. Rev.* **1992**, *45*, 2762–2771. [CrossRef] [PubMed]
19. Hirayama, T.; Holdom, B. Can black holes have Euclidean cores? *Phys. Rev.* **2003**, *68*, 044003. [CrossRef]
20. Corda, C. Black hole quantum spectrum. *Eur. Phys. J.* **2013**, *73*, 2665. [CrossRef]
21. Bekenstein, J.; Schiffer, M. Quantum Limitations on the Storage and Transmission of Information. *Int. J. Mod. Phys.* **1990**, *1*, 355–422. [CrossRef]
22. Bousso, R. The Holographic principle. *Rev. Mod. Phys.* **2002**, *74*, 825–874. [CrossRef]
23. Susskind, L. The world as an hologram. *J. Math. Phys.* **1995**, *36*, 6377–6396. [CrossRef]
24. Maldacena, J.M. The Large N Limit of Superconformal Field Theories and Supergravity. *Adv. Theor. Math. Phys.* **1998**, *2*, 231–252. [CrossRef]
25. Witten, E. Anti De Sitter Space and Holography. *Adv. Theor. Math. Phys.* **1998**, *2*, 253–291. [CrossRef]
26. Dong, X.; Silverstein, E.; Torroba, G. De Sitter holography and entanglement entropy. *J. High Energy Phys.* **2018**, *2018*, 50. [CrossRef]
27. Nikolić, H. Resolving the black-hole information paradox by treating time on an equal footing with space. *Phys. Lett. B* **2009**, *678*, 218–221. [CrossRef]
28. Feleppa, F.; Licata, I.; Corda, C. Hartle-Hawking boundary conditions as Nucleation by de Sitter Vacuum. *Phys. Dark Universe* **2019**, *26*, 100381. [CrossRef]
29. Licata, I.; Fiscaletti, D.; Chiatti, L.; Tamburini, F.; Davide, F. CPT symmetry in cosmology and the Archaic Universe. *Phys. Scr.* **2020**, *95*, 075004. [CrossRef]
30. Licata, I.; Chiatti, L. Event-Based Quantum Mechanics: A Context for the Emergence of Classical Information. *Symmetry* **2019**, *11*, 181. [CrossRef]
31. Vistarini, T. Holographic space and time: Emergent in what sense? *Stud. Hist. Philos. Sci. Part B: Stud. Hist. Philos. Mod. Phys.* **2017**, *59*, 126–135. [CrossRef]
32. Crowther, K. As below, so before: 'synchronic' and 'diachronic' conceptions of spacetime emergence. *Synthese* **2020**, 1–29. [CrossRef]

33. Licata, I. In and Out of the Screen. On some new considerations about localization and delocalization in Archaic Theory, in Beyond Peaceful Coexistence. In *The Emergence of Space, Time and Quantum*; Imperial College Press: London, UK, 2016; pp. 559–577.
34. Kim, H. Classical and quantum wormholes in Einstein-Yang-Mills theory. *Nucl. Phys. B* **1998**, *527*, 342–359. [CrossRef]
35. Maldacena, J.; Susskind, L. Cool horizons for entangled black holes. *Fortsch. Phys.* **2013**, *61*, 781–811. [CrossRef]
36. Susskind, L. Copenhagen vs Everett, Teleportation, and ER = EPR. *Fortschr. Phys.* **2016**, *64*, 551–564. [CrossRef]
37. Cao, C.; Carroll, S.; Michalakis, S. Space from Hilbert Space: Recovering Geometry from Bulk Entanglement. *Phys. Rev.* **2017**, *95*, 024031. [CrossRef]
38. Tamburini, F.; Licata, I. General Relativistic Wormhole Connections from Planck-Scales and the ER = EPR Conjecture. *Entropy* **2019**, *22*, 3. [CrossRef]
39. Bose, S.; Mazumdar, A.; Morley, G.W.; Ulbricht, H.; Toroš, M.; Paternostro, M.; Geraci, A.A.; Barker, P.F.; Kim, M.S.; Milburn, G. Spin Entanglement Witness for Quantum Gravity. *Phys. Rev. Lett.* **2017**, *119*, 240401. [CrossRef]
40. Marletto, C.; Vedral, V. Gravitationally Induced Entanglement between Two Massive Particles is Sufficient Evidence of Quantum Effects in Gravity. *Phys. Rev. Lett.* **2017**, *119*, 240402. [CrossRef]
41. Christodoulou, M.; Rovelli, C. On the possibility of laboratory evidence for quantum superposition of geometries. *Phys. Lett. B* **2019**, *792*, 64–68. [CrossRef]
42. Giacomini, F.; Castro-Ruiz, E.; Brukner, C. Quantum mechanics and the covariance of physical laws in quantum reference frames. *Nat. Commun.* **2019**, *10*, 494. [CrossRef]
43. Chiatti, L.; Licata, I. Particle model from quantum foundations. *Quantum Stud. Math. Found.* **2016**, *4*, 181–204. [CrossRef]

© 2020 by the author. Licensee MDPI, Basel, Switzerland. This article is an open access article distributed under the terms and conditions of the Creative Commons Attribution (CC BY) license (http://creativecommons.org/licenses/by/4.0/).

Article

Quantum Hair on Colliding Black Holes

Lawrence Crowell [1] and Christian Corda [2,3,*]

[1] AIAS, Budapest 1011, Hungary; goldenfieldquaternions@gmail.com
[2] Department of Physics, Faculty of Science, Istanbul University, Istanbul 34134, Turkey
[3] International Institute for Applicable Mathematics and Information Sciences, B.M., Birla Science Centre, Adarshnagar, Hyderabad 500063, India
* Correspondence: cordac.galilei@gmail.com

Received: 22 January 2020; Accepted: 2 March 2020; Published: 5 March 2020

Abstract: Black hole (BH) collisions produce gravitational radiation which is generally thought, in a quantum limit, to be gravitons. The stretched horizon of a black hole contains quantum information, or a form of quantum hair, which is a coalescence of black holes participating in the generation of gravitons. This may be facilitated with a Bohr-like approach to black hole (BH) quantum physics with quasi-normal mode (QNM) approach to BH quantum mechanics. Quantum gravity and quantum hair on event horizons is excited to higher energy in BH coalescence. The near horizon condition for two BHs right before collision is a deformed AdS spacetime. These excited states of BH quantum hair then relax with the production of gravitons. This is then argued to define RT entropy given by quantum hair on the horizons. These qubits of information from a BH coalescence should then appear in gravitational wave (GW) data.

Keywords: colliding black holes; quantum hair; bohr-likr black holes

1. Introduction

Quantum gravitation suffers primarily from an experimental problem. It is common to read critiques that it has gone off into mathematical fantasies, but the real problem is the scale at which such putative physics holds. It is not hard to see that an accelerator with current technology would be a ring encompassing the Milky Way galaxy. Even if we were to use laser physics to accelerate particles the energy of the fields proportional to the frequency could potentially reduce this by a factor of about 10^6 so a Planck mass accelerator would be far smaller; it would encompass the solar system including the Oort cloud out to at least 1 light years. It is also easy to see that a proton-proton collision that produces a quantum black hole (BH) of a few Planck masses would decay into around a mole of daughter particles. The detection and track finding work would be daunting. Such experiments are from a practical perspective nearly impossible. This is independent of whether one is working with string theory or loop variables and related models. It is then best to let nature do the heavy lifting for us. Gravitation is a field with a coupling that scales with the square of mass-energy. Gravitation is only a strong field when lots of mass-energy is concentrated in a small region, such as a BH. The area of the horizon is a measure of maximum entropy any quantity of mass-energy may possess [1], and the change in horizon area with lower and upper bounds in BH thermodynamic a range for gravitational wave production. Gravitational waves produced in BH coalescence contains information concerning the BHs configuration, which is argued here to include quantum hair on the horizons. Quantum hair means the state of a black hole from a single microstate in no-hair theorems. Strominger and Vafa [2] advanced the existence of quantum hair using theory of D-branes and STU string duality. This information appears as gravitational memory, which is found when test masses are not restored to their initial configuration [3]. This information may be used to find data on quantum gravitation. There are three main systems in physics, quantum mechanics (QM), statistical mechanics and general relativity

(GR) along with gauge theory. These three systems connect with each other in certain ways. There is quantum statistical mechanics in the theory of phase transitions, BH thermodynamics connects GR with statistical mechanics, and Hawking-Unruh radiation connects QM to GR as well. These are connections but are incomplete and there has yet to be any general unification or reduction of degrees of freedom. Unification of QM with GR appeared to work well with holography, but now faces an obstruction called the firewall [4]. Hawking proposed that black holes may lose mass through quantum tunneling [5]. Hawking radiation is often thought of as positive and negative energy entangled states where positive energy escapes and negative energy enters the BH. The state which enters the BH effectively removes mass from the same BH and increases the entanglement entropy of the BH through its entanglement with the escaping state. This continues but this entanglement entropy is limited by the Bekenstein bound. In addition, later emitted bosons are entangled with both the black hole and previously emitted bosons. This means a bipartite entanglement is transformed into a tripartite entangled state. This is not a unitary process. This will occur once the BH is at about half its mass at the Page time [6], and it appears the unitary principle (UP) is violated. In order to avoid a violation of UP the equivalence principle (EP) is assumed to be violated with the imposition of a firewall. The unification of QM and GR is still not complete. An elementary approach to unitarity of black holes prior to the Page time is with a Bohr-like approach to BH quantum physics [7–9], which will be discussed in next section. Quantum gravity hair on BHs may be revealed in the collision of two BHs. This quantum gravity hair on horizons will present itself as gravitational memory in a GW. This is presented according to the near horizon condition on Reissnor-Nordstrom BHs, which is $AdS_2 \times \mathbb{S}^2$, which leads to conformal structures and complementarity principle between GR and QM.

2. Bohr-Like Approach to Black Hole Quantum Physics

At the present time, there is a large agreement, among researchers in quantum gravity, that BHs should be highly excited states representing the fundamental bricks of the yet unknown theory of quantum gravitation [7–9]. This is parallel to quantum mechanics of atoms. In the 1920s the founding fathers of quantum mechanics considered atoms as being the fundamental bricks of their new theory. The analogy permits one to argue that BHs could have a discrete energy spectrum [7–9]. In fact, by assuming the BH should be the nucleus the "gravitational atom", then, a quite natural question is—What are the "electrons"? In a recent approach, which involves various papers (see References [7–9] and references within), this important question obtained an intriguing answer. The BH quasi-normal modes (QNMs) (i.e., the horizon's oscillations in a semi-classical approach) triggered by captures of external particles and by emissions of Hawking quanta, represent the "electrons" of the BH which is seen as being a gravitational hydrogen atom [7–9]. In References [7–9] it has been indeed shown that, in the the semi-classical approximation, which means for large values of the BH principal quantum number n, the evaporating Schwarzschild BH can be considered as the gravitational analogous of the historical, semi-classical hydrogen atom, introduced by Niels Bohr in 1913 [10,11]. Thus, BH QNMs are interpreted as the BH electron-like states, which can jump from a quantum level to another one. One can also identify the energy shells of this gravitational hydrogen atom as the absolute values of the quasi-normal frequencies [7–9]. Within the semi-classical approximation of this Bohr-like approach, unitarity holds in BH evaporation. This is because the time evolution of the Bohr-like BH is governed by a time-dependent Schrodinger equation [8,9]. In addition, subsequent emissions of Hawking quanta [5] are entangled with the QNMs (the BH electron states) [8,9]. Various results of BH quantum physics are consistent with the results of [8,9], starting from the famous result of Bekenstein on the area quantization [12]. Recently, this Bohr-like approach to BH quantum physics has been also generalized to the Large AdS BHs, see Reference [13]. For the sake of simplicity, in this Section we will use Planck units ($G = c = k_B = \hbar = \frac{1}{4\pi\epsilon_0} = 1$). Assuming that M is the initial BH mass and that E_n is the total energy emitted by the BH when the same BH is excited at the level n in units of Planck mass (then

$M_p = 1$), one gets that a discrete amount of energy is radiated by the BH in a quantum jump in terms of energy difference between two quantum levels [7–9]

$$\Delta E_{n_1 \to n_2} \equiv E_{n_2} - E_{n_1} = M_{n_1} - M_{n_2}$$
$$= \sqrt{M^2 - \tfrac{n_1}{2}} - \sqrt{M^2 - \tfrac{n_2}{2}}, \tag{1}$$

This equation governs the energy transition between two generic, allowed levels n_1 and $n_2 > n_1$ and consists in the emission of a particle with a frequency $\Delta E_{n_1 \to n_2}$ [7–9]. The quantity M_n in Equation (1), represents the residual mass of the BH which is now excited at the level n. It is exactly the original BH mass minus the total energy emitted when the BH is excited at the level n [8,9]. Then, $M_n = M - E_n$, and one sees that the energy transition between the two generic allowed levels depends only on the two different values of the BH principal quantum number and on the initial BH mass [7–9]. An analogous equation works also in the case of an absorption, See References [7–9] for details. In the analysis of Bohr [10,11], electrons can only lose and gain energy during quantum jumps among various allowed energy shells. In each jump, the hydrogen atom can absorb or emit radiation and the energy difference between the two involved quantum levels is given by the Planck relation (in standard units) $E = h\nu$. In the BH case, the BH QNMs can gain or lose energy by quantum jumps from one allowed energy shell to another by absorbing or emitting radiation (Hawking quanta). The following intriguing remark finalizes the analogy between the current BH analysis and Bohr's hydrogen atom. The interpretation of Equation (1) is the energy states of a particle, that is the electron of the gravitational atom, which is quantized on a circle of length [7–9]

$$L = 4\pi \left(M + \sqrt{M^2 - \tfrac{n}{2}} \right). \tag{2}$$

Hence, one really finds the analogous of the electron traveling in circular orbits around the nucleus in Bohr's hydrogen atom. One sees that it is also

$$M_n = \sqrt{M^2 - \tfrac{n}{2}}. \tag{3}$$

Thus the uncertainty in a clock measuring a time t becomes, with the Planck mass is equal to 1 in Planck units,

$$\frac{\delta t}{t} = \frac{1}{2M_n} = \frac{1}{\sqrt{M^2 - \tfrac{n}{2}}}, \tag{4}$$

which means that the accuracy of the clock required to record physics at the horizon depends on the BH excited state, which corresponds to the number of Planck masses it has. More in general, from the Bohr-like approach to BH quantum physics it emerges that BHs seem to be well defined quantum mechanical systems, having ordered, discrete quantum spectra. This issue appears consistent with the unitarity of the underlying quantum gravity theory and with the idea that information should come out in BH evaporation, in agreement with a known result of Page [6]. For the sake of completeness and of correctness, we stress that the topic of this Section, that is, the Bohr-like treatment of BH quantum physics, is not new. A similar approach was used by Bekenstein in 1997 [14] and by Chandrasekhar in 1998 [15].

3. Near Horizon Spacetime and Collision of Black Holes

This paper proposes how the quantum basis of black holes may be detected in gravitational radiation. Signatures of quantum modes may exist in gravitational radiation. Gravitational memory or BMS symmetries are one way in which quantum hair associated with a black hole may be detected [16]. Conservation of quantum information suggests that quantum states on the horizon may be emitted or entangled with gravitational radiation and its quantum numbers and information. In what follows a toy model is presented where a black hole coalescence excites quantum hair on the stretched horizon in the events leading up to the merger of the two horizons. The model is the Poincare disk for spatial surface in time. To motivate this we look at the near horizon condition for a near extremal black hole. The Reissnor-Nordstrom (RN) metric is

$$ds^2 = -\left(1 - \frac{2m}{r} + \frac{Q^2}{r^2}\right)dt^2 + \left(1 - \frac{2m}{r} + \frac{Q^2}{r^2}\right)^{-1}dr^2 + r^2 d\Omega^2.$$

Here Q is an electric or Yang-Mills charge and m is the BH mass. In previous section, considering the Schwarzschild BH, we labeled the BH mass as M instead. The accelerated observer near the horizon has a constant radial distance. For the sake of completeness, we recall that the Bohr-like approach to BH quantum physics has been also partially developed for the Reissnor-Nordstrom black hole (RNBH) in Reference [14]. In that case, the expression of the energy levels of the RNBH is a bit more complicated than the expression of the energy levels of the Schwarzschild BH, being given by (in Planck units and for small values of Q) [14]

$$E_n \simeq m - \sqrt{m^2 + \frac{q^2}{2} - Qq - \frac{n}{2}}, \qquad (5)$$

where q is the total charge that has been loss by the BH excited at the level n. Now consider

$$\rho = \int_{r_+}^{r} dr \sqrt{g_{rr}} = \int_{r_+}^{r} \frac{dr}{\sqrt{1 - 2m/r + Q^2/r^2}}$$

with lower integration limit r_+ is some small distance from the horizon and the upper limit r removed from the black hole. The result is

$$\rho = m \log[\sqrt{r^2 - 2mr + Q^2} + r - m] + \sqrt{r^2 - 2mr + Q^2}\Big|_{r_+}^{r}$$

with a change of variables $\rho = \rho(r)$ the metric is

$$ds^2 = \left(\frac{\rho}{m}\right)^2 dt^2 - \left(\frac{m}{\rho}\right)^2 d\rho^2 - m^2 d\Omega^2, \qquad (6)$$

where on the horizon $\rho \to r$. This is the metric for $AdS_2 \times \mathbb{S}^2$ for AdS_2 in the (t, ρ) variables tensored with a two-sphere $\mathbb{S}^2$ of constant radius $= m$ in the angular variables at every point of AdS_2. This metric was derived by Carroll, Johnson and Randall [17]. In Section 4 it is shown this hyperbolic dynamics for fields on the horizon of coalescing BHs is excited. This by the Einstein field equation will generate gravitational waves, or gravitons in some quantum limit not completely understood. This GW information produced by BH collisions will reach the outside world highly red shifted by the tortoise coordinate $r^* = r' - r - 2m \ln|1 - 2m/r|$. For a 30 solar mass BH, which is mass of some of the BHs which produce gravitational waves detected by LIGO, the wavelength of this ripple, as measured from the horizon to $\delta r \sim \lambda$

$$\delta r' = \lambda - 2m \ln\left(\frac{\lambda}{2m}\right) \simeq 2 \times 10^6 m.$$

A ripple in spacetime originating an atomic distance 10^{-10} m from the horizon gives a $\nu = 150$ Hz signal, detectable by LIGO [18]. Similarly, a ripple 10^{-13} to 10^{-17} cm from the horizon will give a 10^{-1} Hz signal detectable by the eLISA interferometer system [19]. Thus, quantum hair associated with QCD and electroweak interactions that produce GWs could be detected. More exact calculations are obviously required. Following Reference [20], one can use Hawking's periodicity argument from the RN metric in order to obtain an "effective" RN metric which takes into account the BH dynamical geometry due to the subsequent emissions of Hawking quanta. Hawking radiation is generated by a tunneling of quantum hair to the exterior, or equivalently by the reduction in the number of quantum modes of the BH. This process should then be associated with the generation of a gravitational wave. This would be a more complete dynamical description of the response spacetime has to Hawking radiation, just as with what follows with the converse absorption of mass or black hole coalescence. This will be discussed in a subsequent paper. These weak gravitons produced by BH hair would manifest themselves in gravitational memory. The Bondi-Metzner-Sachs (BMS) symmetry of gravitational radiation results in the displacement of test masses [21]. This displacement requires an interferometer with free floating mirrors, such as what will be available with the eLISA system. The BMS symmetry is a record of YM charges or potentials on the horizon converted into gravitational information. The BMS metric provide phenomenology for YM gauge fields, entanglements of states on horizons and gravitational radiation. The physics is correspondence between YM gauge fields and gravitation. The BHs coalescence is a process which converts qubits on the BHs horizons into gravitons. Two BHs close to coalescence define a region between their horizons with a vacuum similar to that in a Casimir experiment. The two horizons have quantum hair that forms a type of holographic "charge" that performs work on spacetime as the region contracts. The quantum hair on the stretched horizon is raised into excited states. The ansatz is made that $AdS_2 \times \mathbb{S}^2$ for two nearly merged BHs is mapped into a deformed AdS_4 for a small region of space between two event horizons of nearly merged BHs. The deformation is because the conformal hyperbolic disk is mapped into a strip. In one dimension lower, the spatial region is a two dimensional hyperbolic strip mapped from a Poincare disk with the same $SL(2, \mathbb{R})$ symmetry. The manifold with genus g for charges has Euler characteristic $\chi = 2g - 2$ and with the 3 dimensions of $SL(2, \mathbb{R})$ this is the index $6g - 6$ for Teichmuller space [21]. The $SL(2, \mathbb{R})$ is the symmetry of the spatial region with local charges modeled as a $U(1)$ field theory on an AdS_3. The Poincare disk is then transformed into $\mathbb{H}_p^2$ that is a strip. The $\mathbb{H}_p^2 \subset AdS_3$ is simply a Poincare disk in complex variables then mapped into a strip with two boundaries that define the region between the two event horizons.

4. *AdS* Geometry in BH Coalescence

The near horizon condition for a near extremal black hole approximates $AdS_2 \times \mathbb{S}^2$. In Reference [17] the extremal blackhole replaces the spacelike region in (r_+, r_-) with $AdS_2 \times \mathbb{S}^2$. For two black holes in near coalescence there are two horizons, that geodesics terminate on. The region between the horizons is a form of Kasner spacetime with an anisotropy in dynamics between the radial direction and on a plane normal to the radial direction. In the appendix it is shown this is for a short time period approximately an AdS_4 spacetime. The spatial surface is a three-dimensional Poincare strip, or a three-dimensional region with hyperbolic arcs. This may be mapped into a hyperbolic space H^3. This is a further correlation between anti-de Sitter spacetimes and black holes, such as seen in AdS/BH correspondences [22]. The region between two event horizons is argued to be approximately AdS_4 by first considering the two BHs separated by some distance. There is an expansion of the area of the $\mathbb{S}^2$ that is then employed with the $AdS_2 \times \mathbb{S}^2$. We then make some estimates on the near horizon condition for black holes very close to merging. To start consider the case of two equal mass black holes in a circular orbit around a central point. We consider the metric near the center of mass $r = 0$ and the distance between the two black holes $d \gg 2m$. In doing this we may get suggestions om how to model the small region between two black holes about to coalesce. An approximate metric for two distant black holes is of the form

$$ds^2 = \left(1 - \frac{2m}{|r+d|} - \frac{2m}{|r+d|}\right) dt^2 - \left(1 - \frac{2m}{|r+d|} - \frac{2m}{|r+d|}\right)^{-1} dr^2 - r^2(d\theta^2 + sin^2\theta d\Phi^2),$$

where $d\Phi = d\phi + \omega dt$, for ω the angular velocity of the two black holes around $r = 0$. With the approximation for a moderate Keplerian orbit we may then write this metric as This metric is approximated with the binomial expansion to $O(r^2)$ and $O(\omega)$ as

$$ds^2 = \left(1 - \frac{2m}{d}\left(1 + 2\frac{r^2}{d^2}\right)\right)dt^2 - \left(1 - \frac{2m}{d}\left(1 + 2\frac{r^2}{d^2}\right)\right)^{-1}dr^2$$
$$- 2r^2\omega\sin\theta d\phi dt - r^2(d\theta^2 + \sin^2\theta d\phi^2).$$

g_{tt} is similar to the AdS_2 g_{tt} metric term plus constant terms and and similarly g_{rr}. It is important to note this approximate metric has expanded the measure of the angular portion of the metric. This means the 2-sphere with these angle measures has more "area" than before from the contribution of angular momentum.

The Ricci curvatures are

$$R_{tt} = R_{rr} \simeq -\frac{4m}{d}, \quad R_{t\phi} \simeq \left[4\left(1 + \frac{4m}{d}\right) + \frac{16mr^2}{d^3}\right]\omega\sin^2\theta,$$

$$R_{\phi\phi} = g_{t\phi}g^{tt}R_{t\phi} \simeq -8r^2\omega^2\sin^4\theta + O\left(\frac{\omega^2}{d}\right), \quad R_{\theta\theta} = 0,$$

where $O(d^{-2})$ terms and higher are dropped. The R_{rr} and $R_{t\phi}$ Ricci curvature are negative and $R_{t\phi}$ positive. The 2-surface in r, ϕ coordinates has hyperbolic properties. This means we have at least the embedding of a deformed version of AdS_3 in this spacetime. This exercise expands the boundary of the disk $\mathbb{D}^2$, in a 2-spacial subsurface, with boundary around each radial distance so there is an excess angle or "wedge" that gives hyperbolic geometry.

The (t, ϕ) curvature components comes from the Riemannian curvature $R_{r\phi tr} = -\frac{1}{2}\omega\alpha^{-1}$ and its contribution to the geodesic deviation equation along the radial direction is

$$\frac{d^2r}{ds^2} + R^r_{\phi tr}U^tU^\phi r = 0$$

or that for $U^t \simeq 1$ and $U^\phi \simeq \omega$

$$\frac{d^2r}{dt^2} \simeq \frac{1}{2}\omega^2 r.$$

This has a hyperbolic solution $r = r_0\cosh(\frac{1}{\sqrt{2}}\omega t)$. The U^ϕ will have higher order terms that may be computed in the dynamics for ϕ Similarly the geodesic deviation equation for ϕ is

$$\frac{d^2\phi}{ds^2} + R^\phi_{rtr}U^tU^r r = 0$$

or cryptically

$$\frac{d^2\phi}{dt^2} \simeq \mathbf{Riem}\ A\cosh(\alpha t)\sinh(\alpha t),$$

for $\mathbf{Riem} \to \mathbf{R}^{\oe}_{rtr}$. This has an approximately linear form for small t that turns around into exponential or hyperbolic forms for larger time. The spatial manifold in the (r, ϕ) variables then have some hyperbolic structure.

It is worth a comment on the existence of Ricci curvatures for this spacetime. The Schwarzschild metric has no Ricci curvature as a vacuum solution. This 2-black hole solution however is not exactly integrable and so mass-energy is not localizable. This means there is an effective source of curvature due to the nonlocalizable nature of mass-energy for this metric. This argument is made in order to justify the ansatz the spacetime between two close event horizons prior to coalescence is AdS_4. Since most of the analysis of quantum field is in one dimension lower it is evident there is a subspace

AdS_3. This is however followed up by looking at geometry just prior to coalescence where the $\mathbb{S}^2$ has more area than it can bound in a volume. This leads to hyperbolic geometry. Above we argue there is an expansion of a disk boundary $\partial\mathbb{D}^2$, and thus hyperbolic geometry. It is then assume this carries to one additional dimension as well. Now move to examine two black holes with their horizons very close. Consider a modification of the $AdS_2 \times \mathbb{S}^2$ metric with the inclusion of more "œarea" in the $\mathbb{S}^2$ portion. The addition of area to $\mathbb{S}^2$ is then included in the metric. In this fashion the influence of the second horizon is approximated by a change in the metric of $\mathbb{S}^2$. The metric is then a modified form of the near horizon metric for a single black hole,

$$ds^2 = \left(\frac{r}{R}\right)^2 dt^2 - \left(\frac{R}{r}\right)^2 dr^2 - (r^2 + \rho^2)d\Omega^2.$$

The term ρ means there is additional area to the $\mathbb{S}^2$ making it hyperbolic. The Riemann curvatures for this metric are:

$$R_{trtr} = -\frac{1}{r^2} - \frac{2\rho^2}{r^2(r^2 + \rho^2)}, \quad R_{r\theta r\theta} = -\frac{\rho^2}{r^2 + \rho^2}, \quad R_{r\phi r\phi} = -\frac{\rho^2}{r^2 + \rho^2}\sin^2\theta, \quad R_{\theta\phi\theta\phi} = \rho^2\sin^2\theta$$

From these the Ricci curvatures are

$$R_{rr} = -\frac{1}{r^2} - 2\frac{\rho^2}{r^2 + \rho^2}^2, \quad R_{\theta\theta} = R_{\phi\phi} = -\left(1 + \frac{R^2}{r^2}\right)\frac{\rho^2}{r^2 + \rho^2}$$

are negative for small values of r. For $r \to 0$ all Ricci curvatures diverge **Ric** $\to -\infty$. The R_{rr} diverges more rapidly, which gives this spacetime region some properties similar to a Kasner metric. However, $R_{rr} - R_{\theta\theta}$ is finite for $r \to \infty$. This metric then has properties of a deformed AdS_4. With the treatment of quantum fields between two close horizons before coalescence the hyperbolic space $\mathbb{H}^2$ is considered as the spatial surface in a highly deformed AdS_3. A Poincare disk is mapped into a hyperbolic strip.

The remaining discussion will now center around the spatial hyperbolic spatial surface. In particular the spatial dimensions are reduced by one. This is then a BTZ-like analysis of the near horizon condition. The 2 dimensional spatial surface will exhibit hyperbolic dynamics for particle fields and this is then a model for the near horizon hair that occurs with the two black holes in this region.

For the sake of simplicity now reduce the dimensions and consider AdS_3 in 2 plus 1 spacetime. The near horizon condition for a near extremal black hole in 4 dimensions is considered for the BTZ black hole. This AdS_3 spacetime is then a foliations of hyperbolic spatial surfaces H^2 in time. These surfaces under conformal mapping are a Poincare disk. The motion of a particle on this disk are arcs that reach the conformal boundary as $t \to \infty$. This is then the spatial region we consider the dynamics of a quantum particle. This particle we start out treating as a Dirac particle, but the spinor field we then largely ignore by taking the square of the Dirac equation to get a Klein-Gordon wave. Define the z and $\bar{z}$ of the Poincare disk with the metric

$$ds^2_{p-disk} = R^2 g_{z\bar{z}} dz d\bar{z} = R^2 \frac{dz d\bar{z}}{1 - z\bar{z}}$$

with constant negative Gaussian curvature $\mathcal{R} = -4/R^2$. This metric $g_{z\bar{z}} = R^2/(1 - \bar{z}z)$ is invariant under the $SL(2, \mathbb{R}) \sim SU(1,1)$ group action, which, for $g \in SU(1,1)$, takes the form

$$z \to gz = \frac{az + b}{\bar{b}z + \bar{a}}, \quad g = \begin{pmatrix} a & b \\ \bar{b} & \bar{a} \end{pmatrix}. \tag{7}$$

The Dirac equation $i\gamma^\mu D_\mu \psi + m\psi = 0$, $D_\mu = \partial_\mu + iA_\mu$ on the Poincare disk has the Hamiltonian matrix

$$\mathcal{H} = \begin{pmatrix} m & H_w \\ H_w^* & -m \end{pmatrix} \tag{8}$$

for the Weyl Hamiltonians

$$H_w = \frac{1}{\sqrt{g_{z\bar{z}}}} \alpha_z \left(2D_z + \frac{1}{2}\partial_z(\ln g_{z\bar{z}}) \right),$$

$$H_w^* = \frac{1}{\sqrt{g_{z\bar{z}}}} \alpha_{\bar{z}} \left(2D_{\bar{z}} + \frac{1}{2}\partial_{\bar{z}}(\ln g_{z\bar{z}}) \right),$$

with $D_z = \partial_z + iA_z$ and $D_{\bar{z}} = \partial_{\bar{z}} + iA_{\bar{z}}$. here α_z and $\tilde{\alpha}_z$ are the 2×2 Weyl matrices. Now consider gauge fields, in this case magnetic fields, in the disk. These magnetic fields are topological in the sense of the Dirac monopole with vanishing Ahranov-Bohm phase. The vector potential for this field is

$$A^\phi = -i\frac{\phi}{2}\left(\frac{dz}{z} - \frac{d\bar{z}}{\bar{z}}\right).$$

the magnetic field is evaluated as a line integral around the solenoid opening, which is zero, but the Stokes' rule indicates this field will be $\phi(\bar{z}-z)/r^2$, for $r^2 = \bar{z}z$. A constant magnetic field dependent upon the volume $\mathbf{V} = \frac{1}{2} dz \wedge d\bar{z}$ in the space with constant Gaussian curvature $\mathcal{R} = -4/R^2$

$$\mathbf{A}^v = i\frac{BR^2}{4}\left(\frac{zd\bar{z} - \bar{z}dz}{1 - \bar{z}z}\right).$$

The Weyl Hamiltonians are then

$$H_w = \frac{1-r^2}{R}e^{-i\theta}\left(\alpha_z\left(\partial_r - \frac{i}{r}\partial_\theta - \frac{\sqrt{\ell(\ell+1)} + \phi}{r} + i\frac{kr}{1-r^2}\right)\right)$$

$$H_w^* = \frac{1-r^2}{R}e^{i\theta}\left(\alpha_{\bar{z}}\left(\partial_r - \frac{i}{r}\partial_\theta + \frac{\sqrt{\ell(\ell+1)} + \phi}{r} + i\frac{kr}{1-r^2}\right)\right), \tag{9}$$

for $k = BR^2/4$. With the approximation that $r \ll 1$ or small orbits the product gives the Klein-Gordon equation

$$\partial_t^2 \psi = R^{-2}\left(\partial_r^2 + \frac{\ell(\ell+1) + \phi^2}{r^2} + k^2 r^2 + (\ell(\ell+1) + \phi^2)k\right)\psi.$$

For $\ell(\ell+1) + \phi^2 = 0$ this gives the Weber equation with parabolic cylinder functions for solutions. The last term $(\ell(\ell+1) + \phi^2)k$ can be absorbed into the constant phase $\psi(r,t) = \psi(r)e^{-it\sqrt{E^2 + \ell(\ell+1) + \phi^2}}$. This dynamics for a particle in a Poincare disk is used to model the same dynamics for a particle in a region bounded by the event horizons of a black hole. With AdS black hole correspondence the field content of the AdS boundary is the same as the horizon of a black hole. An elementary way to accomplish this is to map the Poincare disk into a strip. The boundaries of the strip then play the role of the event horizons. The fields of interest between the horizons are assumed to have orbits or dynamics not close to the horizons. The map is $z = \tanh(\zeta)$. The Klein-Gordon equation is then

$$\partial_t^2 \psi = R^{-2}\left((1 + 2\zeta^2)\partial_\zeta \partial_{\bar{\zeta}} + \frac{\ell(\ell+1) + \phi^2}{|\zeta|^2} - k|\zeta|^2\right)\psi, \tag{10}$$

where the ζ^2 is set to zero under this approximation. The Klein-Gordon equation is identical to the above.

The solution to this differential equation for $\Phi = \ell(\ell+1) + \phi^2$ is

$$\psi = (2\zeta)^{\frac{1}{4}(\sqrt{1-4\Phi}+1)} e^{-\frac{1}{2}k\zeta^2} \times$$

$$\left[c_1 U\left(\frac{1}{4}\left(\frac{E^2R^2}{k} + \sqrt{1-4\Phi}+1\right), \frac{1}{2}(\sqrt{1-4\Phi}+1), k\zeta^2\right) + c_2 L_{\frac{E^2R^2}{k}+\sqrt{1-4\Phi}}^{\frac{1}{2}\sqrt{1-4\Phi}}(k\zeta^2)\right].$$

The first of these is the confluent hypergeometric function of the second kind. For $\Phi = 0$ this reduces to the parabolic cylinder function. The second term is the associated Laguerre polynomial. The wave determined by the parabolic cylinder function and the radial hydrogen-like function have eigenmodes of the form in the diagram above. The parabolic cylinder function $D_n = 2^{n/2}e^{-x^2/4}H_n(x/\sqrt{2})$ with integer n gives the Hermite polynomial. The recursion formula then gives the modes for the quantum harmonic oscillator. The generalized Laguerre polynomial $L_{n-\ell-1}^{2\ell+1}(r)$ of degree $n - \ell - 1$ gives the radial solutions to the hydrogen atom. The associated Laguerre polynomial with general non-integer indices has degree associated with angular momentum and the magnetic fields. This means a part of this function is similar to the quantum harmonic oscillator and the hydrogen atom. The two parts in a general solution have amplitudes c_1 and c_2 and quantum states in between the close horizons of coalescing black holes are then in some superposition of these types of quantum states (See Figure 1).

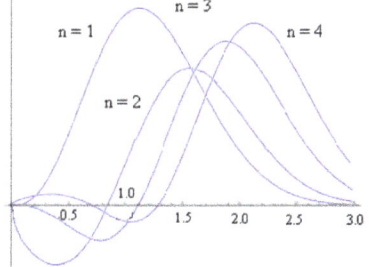
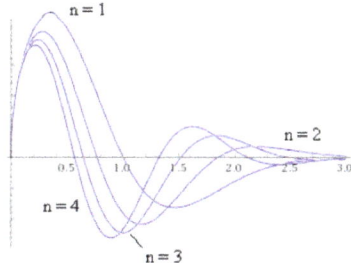

Figure 1. These are the wave function components contributed by the parabolic cylinder functions, or Hermite polynomials and the Laguerre polymomials. These depend on $x^2 = k\zeta^2$ so the wave function is radial. These are not nomalized. (**left**) Solution of the form $x^{1/4}e^{-x^2}H_n(x^2)$ given by parabolic cylinder function for n = 1, 2, 3, ...4 represented as a Hermite polynomial; (**right**) Laguerre wave function $x^{1/4}e^{-x^2}L_n^0(x^2)$ for hydrogen atomic-like states for n = 1,2,3,4.

The Hamiltonian

$$H = \frac{1}{2}|\pi|^2 - \frac{g}{r^2}, \ \pi = -i\partial_r,$$

which contains the monopole field, describes the motion of a gauge particle in the hyperbolic space. In addition, there is a contribution from the constant magnetic field $U = -kr^2/2$. Now convert this theory to a scalar field theory with $r \to \phi$ and $\pi = -i\partial_r\phi$. Finally introduce the dilaton operator D and the scalar theory consists of the operators

$$H_0 = \frac{1}{2}|\pi|^2 - \frac{g}{\phi^2}, \ U = -k\frac{\phi^2}{2}, \ D = \frac{1}{4}(\phi\pi + \pi\phi),$$

where $H_0 + U$ is the field theoretic form of the potential in Equation (9). These potentials then lead to the algebra

$$[H_0, U] = -2iD. \ [H_0, D] = -iH_0, \ [U, D] = iM.$$

This may be written in a more compact form with $L_0 = 2^{-1/2}(H_0 + U)$, which is the total Hamiltonian, and $L_\pm = 2^{-1/2}(U - H_0 \pm iD)$. This leaves the $SL(2, \mathbb{R})$ algebra

$$[L_0, L_\pm] = \pm iL_0, \quad [L_+, L_-] = L_0. \tag{11}$$

This is the standard algebra $\sim \mathfrak{su}(2)$. Given the presence of the dilaton operator this indicates conformal structure. The space and time scale as $(t, x) \to \lambda(t, x)$ and the field transforms as $\phi \to \lambda^\Delta \phi$. The measure of the integral $d^4x \sqrt{g}$ is invariant, where $\lambda = \partial x'/\partial x$ gives the Jacobian $J = det|\frac{\partial x'}{\partial x}|$ that cancels the $\sqrt{g}$ and the measure is independent of scale. In doing this we are anticipating this theory in four dimensions. We then simply have the scaling $\phi \to \lambda^{-1}\phi$ and $\pi \to \pi,$. For the potential term $-g/2\phi^2$ invariance of the action requires $g \to \lambda^{-2}g$ and for $U = -k\frac{\phi^2}{2}$ clearly $k \to \lambda^2 k$. This means we can consider this theory for 2 space plus 1 time and its gauge-like group $SL(2, \mathbb{R})$ as one part of an $SL(2, \mathbb{C}) \sim SL(2, \mathbb{R})^2$. The differential equation number 10 is a modified form of the Weber equation $\psi_{xx} - (\frac{1}{4}x^2 + c)\psi = = 0$ The solution in Abramowit and Stegun are parabolic cylinder functions $D_{-a-1/2}(x)$, written according to hypergeometric functions. The ζ^{-1} part of the differential equation contributes the Laguerre polynomial solution. If we let $\zeta = e^{x/2}$ and expand to quadratic powers we then have the potential in the variable x.

$$V(x) = -(g + k) + \frac{1}{2}(k - g)(x^2 + x^4),$$

for g and k the constants in H_0 and U. The Schrodinger equation for this potential with a stationary phase in time has the parabolic cylinder function solution

$$\psi(x) = c_1 D_{\frac{\beta^2 - 4(\alpha + 2\sqrt{2}\alpha^{3/2})}{16\sqrt{2}\alpha^{3/2}}} \left(\frac{\beta(1 + 4x)}{\sqrt{2}(2\alpha)^{3/4}}\right) + c_2 D_{\frac{-\beta^2 - 4(\alpha - 2\sqrt{2}\alpha^{3/2})}{16\sqrt{2}\alpha^{3/2}}} \left(\frac{i\beta(1 + 4x)}{\sqrt{2}(2\alpha)^{3/4}}\right),$$

where $\alpha = g + k$ and $\beta = k - g$. The parabolic cylinder function describes a theory with criticality, which in this case has with a Ginsburg-Landau potential. The field theory form also has parabolic cylinder function solutions. The field-theory with the field expanded as $\phi = e^\chi$ is expanded around unity so $\phi \simeq 1 + \chi + \frac{1}{2}\chi^2$. A constant C such that $\chi \to C\chi$ is unitless is assumed or implied to exist. The Lagrangian for this theory is

$$\mathcal{L} = \frac{1}{2}\partial_\mu \chi \partial^\mu \chi + \alpha + \frac{1}{2}\mu^2 \chi^2 + 2\beta\mu\chi.$$

The constant μ, standing for mass and absorbing α, is written for dimensional purposes. We then consider the path integral $Z = D[\chi]e^{-iS-iX J}$. Consider the functional differentials acting on the path integral

$$\left((p^2 + m^2)\frac{\delta}{\delta J} - 2i\beta\right) Z = -i\left\langle \frac{\delta S}{\delta \chi}\right\rangle,$$

where $\partial_\mu \chi = p_\mu \chi$. The Dyson-Schwinger theorem tells us that $\langle \frac{\delta S}{\delta \chi}\rangle = \langle J\rangle$ mean we have a polynomial expression $\langle \frac{1}{2}(p^2 + m^2)\chi - i\beta - J\rangle = 0$, where we can trivially let $J - i\beta \to J$. This does not lead to parabolic cylinder functions. There has been a disconnect between the ordinary quantum mechanical theory and the QFT. We may however, continue the expansion to quartic terms. This will also mean there is a cubic term, we may impose that only the real functional variation terms contribute and so only even power of the field define the Lagrangian

$$\mathcal{L} \to \frac{1}{2}\partial_\mu \chi \partial^\mu \chi + \alpha + \frac{1}{2}\mu^2 \chi^2 + \frac{1}{4}\lambda \chi^4,$$

where $\frac{2}{3}\alpha \to \frac{1}{4}\lambda$. The functional derivatives are then

$$\left((p^2 + m^2)\frac{\delta}{\delta J} + \lambda\frac{\delta^3}{\delta J^3}\right) Z = -i\left\langle\frac{\delta S}{\delta\chi}\right\rangle,$$

This cubic form has three parabolic cylinder solutions. We may think of this as $ap + bp^3 = J$ and is a cubic equation for the source J that is annulled at three points. The correspond to distinct solutions with distinct paths. These three solutions correspond to three contours and define three distinct vacua. The overall action is a quartic function, which will have three distinct vacua, where one of these is the low energy physical vacua. It is worth noting this transformation of the problem has converted it into a system similar to the Higgs field. This system with both harmonic oscillator and a Coulomb potentials is conformal and it maps into a system with parabolic cylinder functions solutions. In effect there is a transformation *harmonic oscillator states* ↔ *hydrogen* − *like states*. The three solutions would correspond to the continuance of conformal symmetry, but where the low energy vacuum for one of these may not appear to be conformally invariant. This scale transformation above is easily seen to be the conformal transformation with $\lambda = \Omega$. The scalar tensor theory of gravity for coupling constant $\kappa = 16\pi G$

$$S[g, \phi] = \int d^4x \sqrt{g}\left(\frac{1}{\kappa}R + \frac{1}{2}\partial_\mu\phi\partial^\mu\phi + V(\phi)\right). \tag{12}$$

This then has the conformal transformations

$$g'_{\mu\nu} = \Omega^2 g_{\mu\nu}, \quad \phi' = \Omega^{-1}\phi, \quad \Omega^2 = 1 + \kappa\phi^2.$$

with the transformed action

$$S[g', \phi'] = \int d^4x\sqrt{g'}\left(\frac{1}{\kappa}R' + \frac{1}{2}g'^{\mu\nu}\partial_\mu\phi'\partial_\nu\phi' + V(\phi') + \frac{1}{12}R\phi'^2\right). \tag{13}$$

There is then a hidden $SO(3, 1) \simeq SL(2, \mathbb{C})$ symmetry. Given an internal index on the scalar field ϕ^i there is a linear $SO(n)$ transformation $\delta\phi^i = C^{ijk}\phi_j\delta\tau_k$ for τ_k a parameter. There is also a nonlinear transformation from Equation (12) as $\delta\phi^i = (1 + \kappa\phi^2)^{1/2}\kappa\delta\chi^i$ for χ^i a parameterization. In the primed coordinates the scalar field and metric transform as

$$\delta\phi^i = \delta\tau^i - \kappa\phi'^i\phi^j\delta\chi^j$$

$$\delta g_{\mu\nu} = \frac{2g'_{\mu\nu}\kappa\phi'^i\delta\chi^i}{1 - \kappa\phi'^2}. \tag{14}$$

The gauge-like dynamics have been buried into the scalar field. With this semi-classical model the scalar field adds some renormalizability. Further this model is conformal. The conformal transformation mixes the scalar field, which is by itself renormalizable, with the spacetime metric. Quantum gravitation is however difficult to renormalize. Yet we see the linear group theoretic transformation of the scalar field in $SO(n)$ is nonlinear in $SO(n, 1)$. Conformal symmetry is manifested in sourceless spacetime, or spatial regions without matter or fields. The two dimensional spatial surface in AdS_3 is the Poincare disk that with complexified coordinates has metric with $SL(2, \mathbb{R})$ algebraic structure. This may of course be easily extended into $SL(2, \mathbb{C})$ as $SL(2, \mathbb{R}) \times SL(2, \mathbb{R})$. In this conformal setting quantum states share features similar to the emission of photons by a harmonic oscillator or an atom. The orbits of these paths are contained in regions bounded by hyperbolic surfaces, or arcs for the two dimensional Poincare disk. The entropy associated with these arcs is a measure of the area contained within these curves. This is in a nutshell the Mirzakhani result on entropy for hyperbolic curves. This development is meant to illustrate how radiation from black holes is produced by quantum mechanical means not that different from bosons produced by a harmonic oscillator or atom. Hawking radiation in principle is detected with a wavelength not different from the size of the black hole. The wavelength approximately equal to the Schwarzschild radius has energy $E = h\nu$ corresponding to a unit mass emitted. The mass of the black hole is n of these units and it is easy to find $m_p = \sqrt{\hbar c/G}$. These modes emitted are Planck units of mass-energy that reach $\mathcal{I}^\infty$. In the case of gravitons, these carry gravitational memory. For the coalescence of black holes gravitational waves are ultimately gravitons. For Hawking radiation there is the metric back reaction, which in a quantum mechanical setting is an adjustment of the black hole with the emission of gravitons. The emission of Hawking radiation might then be compared to a black hole quantum emitting a Planck unit of black hole that then decays into bosons. The quantum induced change in the metric is a mechanism for producing gravitons. In the coalescence of black holes the quantum hair on the stretched horizons sets up a type of Casimir effect with the vacuum that generates quanta. In general these are gravitons. We might see this as not that different from a scattering experiment with two Planck mass black holes. These will coalesce, form a larger black hole, produce gravitons, and then quantum states excited by this process will decay. The production of gravitons by this mechanism is affiliated with normal modes in the production of gravitons, which in principle is not different from the production of photons and other particles by other quantum mechanical processes. I fact quantum mechanical processes underlying black hole coalescence might well be compared to nuclear fusion. The 2 LIGOs, plus now the VIRGO detector, are recording and triangulating the positions of distant black hole collisions almost weekly. This information may contain quantum mechanical information associated with quantum gravitation. This information is argued below to contain BMS symmetries or information. This will be most easily detected with a space-based system such as eLISA, where the shift in metric positions of test masses is most readily detectable. However, preliminary data with the gross displacement of the LIGO mass may give preliminary information as well.

5. Discussion

The coalescence of two black holes is a form of scattering. We may think of black holes as an excited state of the quantum gravity field and a sort of elementary particle. The scattering of two black holes results in a larger black hole plus gravitational radiation. This black hole will then emit Hawking radiation. Thus, in general the formation of black holes, their coalescence and ultimate quantum evaporation is an intermediate processes in a general scattering theory.

Quantum hair is a set of quantum fields that build up quantum gravitation, in the manner of gauge-gravity duality and BMS symmetry. This is holography, with the fields on the horizons of two BHs that determine the graviton/GW content of the BH coalescence. A detailed analysis of this may reveal BMS charges that reach $\mathcal{I}^+$ are entangled with Hawking radiation by a form of entanglement swap. In this way Hawking radiation may not be entangled with the black hole and thus not with previously emitted Hawking radiation. This will be addressed later, but a preliminary to this idea is seen in Reference [23], for disentanglement between Hawking radiation and a black hole. The authors are working on current calculations where this is an entanglement swap with gravitons. The black hole production of gravitons in general is then a manifestation of quantum hair entanglement. It is illustrative for physical understanding to consider a linearized form of gravitational memory. Gravitational memory from a physical perspective is the change in the spatial metric of a surface according to Reference [3]

$$\Delta h_{+,\times} = \lim_{t\to\infty} h_{+,\times}(t) - \lim_{t\to-\infty} h_{+,\times}(t).$$

Here, $+$ and $\times$ refer to the two polarization directions of the GW. See Reference [24] for more on this. Quantum hair on two black holes just before coalescence are highly excited and contribute to spacetime curvature, or in a full context of quantum gravitation the generation of gravitons. As yet there is no complete theory of quantum gravity, but it is reasonable to think of gravitational radiation as a classical wave built from many gravitons. Gravitons have two polarizations and a state $|\Psi_{+,\times}\rangle$ the density matrix $\rho_{+,\times} = |\Psi_{+,\times}\rangle\langle\Psi_{+,\times}|$ then defines entropy $S = \rho_{+,\times} log(\rho_{+,\times})$ that with this near horizon condition of *AdS* with a black hole is a form of Mirzakhani entropy measure in hyperbolic space. The gravitons emitted are generated by quantum hair on the colliding black holes. These will contribute to gravitational waves, and in general with BMS translations that bear quantum information from quantum hair.

This theory connects to fundamental research, The entanglement entropy of CFT_2 entropy with AdS_3 lattice spacing a is

$$S \simeq \frac{R}{4G}ln(|\gamma|) = \frac{R}{4G}ln\left[\frac{\ell}{L} + e^{2\rho_c}sin\left(\frac{\pi\ell}{L}\right)\right].$$

where the small lattice cut off avoids the singular condition for $\ell = 0$ or L for $\rho_c = 0$. For the metric in the form $ds^2 = (R/r)^2(-dt^2 + dr^2 + dz^2)$ the geodesic line determines the entropy as the Ryu-Takayanagi (RT) result [25]

$$S = \frac{R}{2G}\int_{2\ell/L}^{\pi/2}\frac{ds}{\sin s} = -\frac{R}{2G}ln[cot(s) + csc(s)]\Big|_{2\ell/L}^{\pi/2}$$

$$\simeq \frac{R}{2G}ln\left(\frac{\ell}{L}\right),$$

which is the small ℓ limit of the above entropy. The RT result specifies entropy, which is connected to action $S_a \leftrightarrow S_e$ [26]. Complexity, a form of Kolmogoroff entropy [27], is $S_a/\pi\hbar$ which can also assume the form of the entropy of a system $S \sim k \, log(dim \, \mathcal{H})$ for $\mathcal{H}$ the Hilbert space and the dimension over the number of states occupied in the Hilbert space. There is also complexity as the volume of the Einstein-Rosen bridge [28] vol/GR_{ads} or equivalently the RT area $\sim vol/R_{AdS}$. There is an equivalency between entropy or complexity according to the geodesic paths in hyperbolic $\mathbb{H}^2$ by geometric means [21]. This should generalize to $\mathbb{H}^3 \subset AdS_4$. The generation of gravitational waves should have an underlying quantum mechanical basis. It is sometimes argued that spacetime physics may not be at all quantum mechanical. This is probably a good approximation for energy sufficient orders of magnitude lower than the Planck scale. However, if we have a scalar field that define the metric $g' = g'(g, \phi)$ with action $S[g, \phi]$ then a quantum field ϕ and a purely classical g means the transformation of g by this field has no quantum physics. In particular for a conformal theory $\Omega = 1 + \kappa\phi^a\phi^a$, here a an internal index, the conformal transformation $g'_{\mu\nu} = \Omega^2 g_{\mu\nu}$ has no quantum content. This is an apparent inconsistency. For the inflationary universe the line element

$$ds'^2 = g'_{\mu\nu}dx^\mu dx^\nu = \Omega^2(du^2 - d\Sigma^{(3)}),$$

with $dt/du = \Omega^2$ gives an FLRW or de Sitter-like line element that expands space with $\Omega^2 = e^{t\sqrt{\Lambda/3}}$. The current slow accelerated universe we observe is approximately of this nature. The inflation scalars are then fields that stretch space as a time dependent conformal transformation and are quantum mechanical. The generation of gravitational waves is ultimately the generation of gravitons. Signatures of these quantum effects in black hole coalescence will entail the measurement of quantum information. Gravitons carry BMS charges and these may be detected with a gravitational wave interferometer capable of measuring the net displacement of a test mass. The black hole hair on the stretched horizon is excited by the merger and these results in the generation of gravitons. The Weyl Hamiltonians in Equation (9) depend on the curvature as $\propto \sqrt{\mathcal{R}}$. For the curvature extreme during the merging of black holes this means many modes are excited. The two black holes are pumped with energy by the collision, this generates or excites more modes on the horizons, where this results in a black hole with a net larger horizon area. This results in a metric response, or equivalently the generation of gravitons. Quantum normal modes are given by independent eigen-states, such as with quantum harmonic oscillator states. The harmonic oscillator states are well known to be given by the Hermite polynomials, which are a special case of parabolic cylinder functions. Rydberg states are also a form of normal modes. The quantum states for the hyperbolic geometry of black hole mergers are a generalization of these forms of states. The excitation of quantum hair in such a merger and the production of gravitons is a converse situation for the emission of Hawking radiation. In both cases there is a dynamical response of the metric, which is associated with gravitons. Currently a "by hand" correction called back reaction is used in models. A more explicit discussion on the production of gravitons is beyond the scope here. However, the parabolic cylinder functions and the Laguerre functions clearly play a role in quantum production of gravitons in BH coalescence. This means quantum gravitation should have signatures of much the same physics as atomic physics or the role of electrons and phonons in solids. The major import of this expository is to propose quantum gravitational signatures in the coalescence of black holes. This would point to quantum hair and the generation of gravitons. This would be a clear signature of quantum gravitation. While there is plenty of further development needed to compute more firm predictions, the generic result is that gravitational waves from colliding black holes have some quantum gravitational signatures. These signatures are to be found in gravitational memory. Further, this long-term adjustment of spacetime metric deviates form a purely classical expected result. With further advances in gravitational wave interferometry, in particular with the future eLISA space mission, it should be possible to detect elements of gravitons and quantum gravitation.

Author Contributions: Investigation, L.C. and C.C. All authors have read and agreed to the published version of the manuscript.

Funding: This research received no external funding.

Acknowledgments: The Authors thank the Editors and the Referees for useful comments

Conflicts of Interest: The authors declare no conflict of interest.

References

1. Bekenstein, J.D. Universal upper bound on the entropy-to-energy ratio for bounded systems. *Phys. Rev. D* **1981**, *23*, 287–298. [CrossRef]
2. Strominger, A.; Vafa, C. Microscopic origin of the Bekenstein-Hawking entropy. *Phys. Lett. B* **1996**, *379*, 99–104. [CrossRef]
3. Favata, M. The gravitational-wave memory effect. *Class. Quantum Gravity* **2010**, *27*, 084036. [CrossRef]
4. Almheiri, A.; Marolf, D.; Polchinski, J.; Sully, J. Black holes: Complementarity or firewalls? *arXiv* **2013**, arXiv:1207.3123.
5. Hawking, S.W. Black hole explosions? *Nature* **1974**, *248*, 30–31. [CrossRef]
6. Page, D.N. Information in Black Hole Radiation. *Phys. Rev. Lett. arXiv* **1993**, arXiv:hep-th/9306083.
7. Corda, C. Precise model of Hawking radiation from the tunnelling mechanism . *Class. Quantum Gravity* **2015**, *32*, 195007. [CrossRef]
8. Corda, C. Time dependent Schrödinger equation for black hole evaporation: No information loss. *Ann. Phys.* **2015**, *353*, 71–82. [CrossRef]
9. Corda, C. Quasi-normal modes: The "electrons" of black holes as "gravitational atoms"? Implications for the black hole information puzzle. *Adv. High Energy Phys.* **2015**, *2015*, 867601. [CrossRef]
10. Bohr, N. On the Constitution of Atoms and Molecules (Part I). *Philos. Mag.* **1913**, *26*, 1–25. [CrossRef]
11. Bohr, N. XXXVII. On the constitution of atoms and molecules. *Philos. Mag.* **1913**, *26*, 476–502. [CrossRef]
12. Bekenstein, J.D. The quantum mass spectrum of the Kerr black hole. *Lett. Al Nuovo Cimento (1971-1985)* **1974**, *11*, 467–470. [CrossRef]
13. Sun, D.Q.; Wang, Z.L.; He, M.; Hu, X.R.; Deng, J.B. Hawking radiation-quasinormal modes correspondence for large AdS black holes. *Adv. High Energy Phys.* **2017**, *2017*, 4817948. [CrossRef]
14. Bekenstein, J. Quantum Black Holes as Atoms. In *Proceedings of the Eight Marcel Grossmann Meeting*; World Scientic: Singapore, 1999; pp. 92–111.
15. Chandrasekhar, S. *The Mathematical Theory of Black Holes*; Reprinted Edition; Oxford University Press: Oxford, UK, 1998; p. 205.

16. Strominger, A.; Zhiboedov, A. Gravitational Memory, BMS Supertranslations and Soft Theorems. *J. High Energy Phys.* **2016**, *2016*, 86. [CrossRef]
17. Carroll, S.M.; Johnson, M.C.; Randall, L. Extremal limits and black hole entropy. *J. High Energy Phys.* **2009**, *2009*, 109. [CrossRef]
18. Moore, C.; Cole, R.; Berry, C. Gravitational Wave Detectors and Sources. Available online: http://gwplotter.com/ (accessed on 1 March 2020).
19. eLISA Interferometer System. Available online: www.lisamission.org (accessed on 1 March 2020).
20. Corda, C. Non-strictly black body spectrum from the tunnelling mechanism. *Ann. Phys.* **2013**, *337*, 49–54. [CrossRef]
21. Eskin, A.; Mirzakhani, M. Counting closed geodesics in Moduli space. *arXiv* **2008**, arXiv:0811.2362.
22. Berman, D.S.; Parikh, M.K. Holography and Rotating AdS Black Holes. *Phys. Lett. B* **1999**, *463*, 168–173. [CrossRef]
23. Hossenfelder, S. Disentangling the Black Hole Vacuum. *Phys. Rev. D* **2015**, *91*, 044015. [CrossRef]
24. Braginsky, V.B.; Thorne, K.S. Gravitational Wave Bursts with Memory and Experimental Prospects. *Nature* **1987**, *327*, 123–125. [CrossRef]
25. Ryu, S.; Takayanagi, T. Aspects of Holographic Entanglement Entropy. *J. High Energy Phys.* **2006**, doi:10.1088/1126-6708/2006/08/045. [CrossRef]
26. Harlow, D. The Ryu-Takayanagi Formula from Quantum Error Correction. *Commun. Math. Phys.* **2017**, *354*, 865–912. [CrossRef]
27. Kolmogorov, A.N. Entropy per unit time as a metric invariant of automorphisms. *Dokl. Russ. Ac. Sci.* **1959**, *124*, 754.
28. Susskind, L. ER=EPR, GHZ, and the Consistency of Quantum Measurements. *arXiv* **2014**, arXiv:1412.8483.

© 2020 by the authors. Licensee MDPI, Basel, Switzerland. This article is an open access article distributed under the terms and conditions of the Creative Commons Attribution (CC BY) license (http://creativecommons.org/licenses/by/4.0/).

Article

General Relativistic Wormhole Connections from Planck-Scales and the ER = EPR Conjecture

Fabrizio Tamburini [1,*,†] **and Ignazio Licata** [2,3,4,†]

1 ZKM—Zentrum für Kunst und Medientechnologie, Lorentzstr. 19, D-76135 Karlsruhe, Germany
2 Institute for Scientific Methodology (ISEM), Via Ugo La Malfa 153, I-90146 Palermo, Italy; ignazio.licata3@gmail.com
3 School of Advanced International Studies on Theoretical and Nonlinear Methodologies of Physics, I-70124 Bari, Italy
4 International Institute for Applicable Mathematics and Information Sciences (IIAMIS), B.M. Birla Science Centre, Adarsh Nagar, Hyderabad 500 463, India
* Correspondence: fabrizio.tamburini@gmail.com
† These authors contributed equally to this work.

Received: 8 November 2019; Accepted: 17 December 2019; Published: 18 December 2019

Abstract: Einstein's equations of general relativity (GR) can describe the connection between events within a given hypervolume of size L larger than the Planck length L_P in terms of wormhole connections where metric fluctuations give rise to an indetermination relationship that involves the Riemann curvature tensor. At low energies (when $L \gg L_P$), these connections behave like an exchange of a virtual graviton with wavelength $\lambda_G = L$ as if gravitation were an emergent physical property. Down to Planck scales, wormholes avoid the gravitational collapse and any superposition of events or space–times become indistinguishable. These properties of Einstein's equations can find connections with the novel picture of quantum gravity (QG) known as the "Einstein–Rosen (ER) = Einstein–Podolski–Rosen (EPR)" (ER = EPR) conjecture proposed by Susskind and Maldacena in Anti-de-Sitter (AdS) space–times in their equivalence with conformal field theories (CFTs). In this scenario, non-traversable wormhole connections of two or more distant events in space–time through Einstein–Rosen (ER) wormholes that are solutions of the equations of GR, are supposed to be equivalent to events connected with non-local Einstein–Podolski–Rosen (EPR) entangled states that instead belong to the language of quantum mechanics. Our findings suggest that if the ER = EPR conjecture is valid, it can be extended to other different types of space–times and that gravity and space–time could be emergent physical quantities if the exchange of a virtual graviton between events can be considered connected by ER wormholes equivalent to entanglement connections.

Keywords: wormholes; entanglement; ER = EPR; relativistic quantum information; Planck scales

1. Introduction

The formulation of an effective theory of quantum gravity can be considered the holy grail of modern physics. Gravitation was the first force to be mathematically described by Newton and it is the last force of nature that has yet to be quantized. As pointed out by DeWitt in his early pioneering works [1–5], since the introduction of quantum field theory around 1930 by Heisenberg, Dirac, Pauli, Fock, Jordan and others, many attempts were made to find a robust and logically closed method of quantizing the gravitational field, without success, even if Einstein's equations are known to remain valid down to the Planck scales. Rosenfeld [6,7] realized the difficulty of finding general methods to quantize gravity and that the quanta of the field, if they do exist, cannot give observational effects until reaching a very high energy $E_p = \sqrt{\hbar c^5/G} \simeq 1.22 \times 10^{19}$ GeV that corresponds to the so-called Planck length, $L_p = \sqrt{\hbar G/c^2}$. Thus, Planck scales were somehow "artificially" introduced in the framework of general relativity (GR), a classical theory, in the attempt of building a quantum theory of gravitation based on the finite quantum of action h linked with the gravitational constant G and the speed of light c.

In principle, following the works by Pauli, De Witt and other pioneers in this field, the fundamental building block for a quantum theory of gravity is the graviton, a spin-2 massless particle with the well-known limitations in the building of a QG theory due to the coupling constant of the gravitational field that depends on the inverse square of the mass. This coupling constant makes the Einstein–Hilbert Lagrangian of quantum gravity divergent at the loop level. It is a non-renormalizable theory, unless introducing additional concepts such as supersymmetry like in string theory and supergravity. Up to now no supersymmetric partners of the known quanta have been found from the Large Hadron Collider and other experiments presented by the Particle Data Group [8]. At the present moment one can consider different approaches, including string theory scenarios that do not require supersymmetric partners at the explored energies, or to consider a model of the Universe without strings, or adopt the approach of loop quantum gravity [9], where gravitons do not represent the building blocks of the theory. The interactions between events that can be ascribed to graviton exchanges can be recovered in a weak field limit approximation. The exchange of a virtual graviton between two particles does not have the support of an actual theory of quantum gravity. As an example, in string theory and in most quantum field theory (QFT) scenarios, in the building of the theory, one must introduce the quanta of the associated field. The quanta are introduced in terms of quantized excitations on a classically fixed background. The main conceptual problem for the formulation of a consistent theory of QG is that this theory must unify and contain as special cases both GR and quantum mechanics (QM). Unfortunately, GR has concepts and mathematical structures that are incompatible with those of QM and vice versa, with the result that the two theories do not communicate between each other. GR is a local deterministic theory based on point-to-point connections of events and observers that define a four-dimensional manifold. Einstein in 1947, his latest memoirs, stated that space–time is made with connections between events and, more precisely, with coincidences of events. On the other hand, QM presents non-locality and the well-known probabilistic behavior from the deterministic equations that rule the quanta.

This contrast between GR and QG can find a fusion in the simple heuristic approach formulated by Susskind and Maldacena who, starting from the quantum mechanical language, set an hypothetical equivalence between non-traversable wormhole connections of two (or more) particles or events in space–time through Einstein–Rosen (ER) bridges and entangled states (the idea that wormholes and flux tubes can play a role in quantum mechanics and quantum field theory is not new, in particular for systems with electric and/or magnetic charges and their renormalization has earlier work in [10,11]), and the quantum properties of the "spooky action at distance" of Einstein–Podolski–Rosen (EPR) states [12–15]. The ER=EPR equivalence was first defined in Anti-de-Sitter (AdS) space–times in their equivalence with CFTs [16–19]. In other words, EPR entangled particles are supposed to be equivalent to connections obtained through ER wormholes involving the concept of entanglement entropy to describe these many-body quantum state/wormhole connections, even if the ER = EPR equivalence is more evident with monogamous entangled pairs [20]. Spacetime is supposed to emerge from quantum entanglement, as discussed in [21] where, from some examples where gauge theory/gravity duality is valid, one finds that the emergence of space–time is related to the quantum entanglement of the degrees of freedom present in these quantum systems. Superpositions of quantum states corresponding to disconnected space–times can give rise to states that are interpreted in terms of classically connected space–times. In this vision, gravity can be also interpreted as an entropic force, a thermodynamic property of physical systems defined in an holographic scenario: gravity and space–time connections are emergent phenomena from the degrees of freedom of a physical system encoded in an holographic boundary or to emerge from a background-free approach by using quantum entanglement [22,23]. At all effects, one can conclude that space–time is built with the quantum information shared between EPR states that are equivalently connected with an ER wormhole.

While ER wormholes are classical solutions of GR, a local deterministic theory, quantum entanglement, is instead one of the most intriguing quantum physical aspects of nature characterized by non-locality and the stochastic properties of quantum mechanics. Entanglement occurs when a pair of particles (or a group of quanta) is generated in a way that the quantum state of each particle of the pair cannot be described independently of the state of the others even if they are separated by large distances. For a deeper insight see [24–29]. In ER = EPR, causality is not violated. ER bridges do not violate causality because of the topological censorship, which forbids ordinary traversable wormholes; EPR states, instead, prevent causality violations because of the properties of entangled states described by Bell's inequalities—no information is transferred between the two entangled states during the wavefunction collapse of the entangled pair as each quantum state in an EPR pair cannot be described independently of the other states [24,26]. The ER = EPR equivalence is valid if there are no traversable wormhole solutions that do not require the violation of the strong and/or weak energy conditions [30–33], they may instead behave as quantum communication channels between the quantum fields there defined [34].

The ER = EPR conjecture was initially formulated in the gravity/gauge theory equivalence between Anti-de Sitter (AdS) space–times and conformal field theories (CFTs) by Maldacena (gauge/gravity duality) within a relationship between the entanglement entropy of a set of black holes and the cross-section area of ER bridges connecting them. AdS space–times represent an elegant solution of Einstein's equations with negative curvature where the outer boundary is a surface and, in the CFT, correspondence quanta can interact and generate the holographic universe there contained. The AdS/CFT correspondence provides a complete non-perturbative definition of gravity with quantum field theory, extending this correspondence also to space–time scenarios of quantum gravity where the asymptotic behavior of the space–time is that of AdS space–time. The AdS/CFT correspondence plays a key role in the calculations of strong coupled quantum field theories. When the boundary theory is strongly coupled, the bulk theory is weakly coupled, and vice versa. A strongly coupled field theory can have an AdS dual gravity description weakly coupled and therefore calculable and vice versa. As an example, if the curvature of the AdS space increases, the gravitational coupling becomes stronger and the boundary coupling is weaker.

To extend ER = EPR conjecture to space–times different from the Anti-de Sitter solution, one has to investigate how much ER = EPR depends strictly on AdS/CFT correspondence and from the properties of wormholes also in de Sitter (dS) space–times. First of all we must consider that AdS/CFT is a structural correspondence between bulk and screen, but does not contain in itself any specific indication of the possible dynamics of wormhole formation. Wormholes require a cosmological scenario. For example, it is plausible that the wormholes were formed in the initial chaotic phases of the universe with a rate similar to that of the formation of mini black holes (BHs), with very specific traces as regards the event horizon, as discussed in [35–37]. This aspect is decisive because all the problems related to the quantum aspect of the wormholes imply a cosmological background capable of providing a plausible scenario for their existence described by the Ryu–Takanayagi's entropy that relates the entanglement entropy in CFT and the geometry of AdS space–times. The Ryu–Takanayagi formula is a generalization of the BH entropy formula by Bekenstein–Hawking [38,39] to a whole class of holographic theories [40,41] where gravitational models with dimension D are dual to a gauge theory in dimension $D-1$.

In these recent years, the interest in the maximum symmetry properties of de Sitter's space gave this structure a new centrality with respect to AdS. One of the most relevant problems was to project a hologram of a quantum particle that lives in the infinite future of AdS, which makes it difficult to describe real-time space in holographic terms. In particular, the main classes of essential results must be mentioned here: the CPT Universe [42] and the numerous results on the non-locality in dS space–times [43–47]. Of relevant importance is the so-called "uplifting" technique by Dong et al. [48] where two Anti-de Sitter space–times are transformed into a de Sitter space–time. The uplifting changes the curvature of two "saddle-shaped" AdS space–times that, once warped, are glued together along their rims and turned into a "bowl-shaped" dS space–time via entanglement or more general two-throated Randall–Sundrum systems [49,50] and the CFTs relative to both hemispheres become coupled with each other. In this way one forms a single quantum system that is holographically dual to the entire spherical de Sitter space, defined on its boundary located at a finite distance away. The technique of uplifting two AdS into a dS permits us to modify the curvature in a more general way than that offered by the set of local transformations obtained through Wick rotations that can only act locally—the curvature changes everywhere by introducing extra fields whose energy density acts as an extra source of curvature to landscape the AdS space–time into a dS one. The cosmological constant in the bulk space is then transformed from negative to positive and the holographic projection of the space–time into its boundaries is changed. Some examples are reported by Silverstein and Polchinski [51] or in Vasiliev's higher-spin gravity—in AdS, the boundary theory is an O(N)-vector field theory, while in dS space it becomes an Sp(N) scalar field theory, where N is the number of the vector (scalar) fields of the boundary theories [52,53].

In this work, we analyze Einstein's equations in a finite volume of space–time down to the Planck scale, finding wormhole connections that avoid the singularity problem and an indetermination relationship that involves the Riemann curvature. This finds application to the ER = EPR conjecture—in this framework, geometry behaves as a geodesic tensor network that defines the quantum state properties of a fundamental quantum state of a given metric [54] and a virtual graviton exchange becomes equivalent to entanglement to which one can apply the concept of Penrose's decoherence of a quantum state [55]. In this crossing between locality of GR and the emergence of non-locality of QM as in [56,57], where de Sitter space–time is taken as the geometric structure of vacuum, the analysis of Einstein's equations can provide an additional support to the ER = EPR conjecture extended from AdS to dS [58] and to locally Euclidean space–times. This can be interpreted as the route to ER = EPR from general relativity.

2. Wormhole Connections down to Planck Scales from Einstein's Equations

In the ER = EPR scenario, wormhole connections are fundamental in the building of space–time. Consider an entangled quantum system. The emergence of space–time in terms of ER connections, in the gravity picture, is intimately related to the quantum entanglement of degrees of freedom in the corresponding conventional quantum system, building up space–time with quantum entanglement. The ER = EPR equivalence suggests that space–time and gravity may emerge from the degrees of freedom of the field theory. On the other hand, space–time becomes the optimal way to build entanglement starting from wormhole connections. At Planck scales, the Planck area is defined as the area by which the surface of a Schwarzschild black hole increases when in the black hole is injected one bit of information. In a Riemannian manifold (M, g) the scalar curvature in an $(n-1)$ hyperplane relates GR with the entanglement of quantum states in an arbitrary Hilbert space without reference to AdS = CFT or any other holographic boundary construction.

2.1. Einstein's Equations in the Neighborhood of an Event

Einstein's equations are the core of GR—they describe gravity in terms of the curvature of space–time. Spacetime geometry and the metric tensor g_{ik} are determined from Einstein's equations, given the distribution of energy, mass and momentum in space–time encoded in the stress–energy tensor T_{ik}.

In our approach, by assuming that Einstein's equations remain valid down to the Planck scale, we find that the connection between events are achieved through wormhole connections, avoiding the gravitational collapse and the presence of singularities at Planck scales. To this aim, we adopt the approach by Schwinger in the analysis of classical fields [59]—to determine the properties of a field, one cannot measure the field in a point, otherwise, because of the equivalence principle, one finds only a local Minkowskian space–time tangent to the manifold (M, g) in the given point event.

Following Schwinger, from his studies on electromagnetic theory [59,60], the analysis of a classical field must be made in a neighborhood of the event. This approach is clearly valid in electromagnetism and antenna theory: when sensing the electromagnetic field with an antenna, one cannot measure in a point the fluctuations of the electric field. The antenna must have a finite length in space and the field must be measured in a finite time interval to be revealed. As happens for any antenna, as well as for the gravitational field, one has to consider a finite length or a finite hypervolume in which to determine the properties of the field. For the same reasons, one cannot deduce the properties of the gravitational field in a single point and at a given time because of the equivalence principle, as discussed in the free-falling particle paradox [61].

Let us find the main properties of the gravitational field when a finite length L of measure is fixed down to the Planck scale. Given a Riemannian manifold (M, g), where M is the manifold and g the metric tensor, $g \in \otimes^2 \dot{T}$ of tensorial order 2, (written as $g_{(2)}$), Einstein's equations are

$$R_{ik} - \frac{1}{2} R g_{ik} + \Lambda g_{ik} = 8\pi T_{ik} \quad (1)$$

where Λ is the cosmological constant and $R_{ik} = g^{lm} R_{lmik}$ and $R = g^{ik} g^{lm} R_{lmik}$ represent the tensorial and scalar space–time curvature terms obtained from the Riemann tensor R_{iklm} that, in the tensorial index notation, takes the usual well-known form [62]

$$R_{iklm} = \frac{1}{2} \left(\frac{\partial^2 g_{lm}}{\partial x^k \partial x^l} + \frac{\partial^2 g_{kl}}{\partial x^i \partial x^m} - \frac{\partial^2 g_{il}}{\partial x^k \partial x^m} - \frac{\partial^2 g_{km}}{\partial x^i \partial x^l} \right) + g_{np} \left(\Gamma^n_{kl} \Gamma^p_{im} - \Gamma^n_{km} \Gamma^p_{il} \right) \quad (2)$$

the tensor is an element of the rank-four tensors $R_{iklm} \in \otimes^4 \dot{T}$ in the cotangent bundle $\dot{T}$ of the manifold (M, g) that we will indicate with the symbol $R_{(4)}$, where 4 is the tensorial index.

The gravitational field has the fundamental property that, any body, independently from their mass, moves in the same way. This is described by the strong equivalence principle, which suggests that gravity is a geometrical quantity and one cannot measure the gravitational field in an event, as the field becomes locally Galilean and diagonalizable, and the energy of the field cannot be uniquely defined.

In a neighborhood of a given event, space–time is built by chains of events and observers. The building of space–time is obtained by causally transferring information encoded locally in one event of the field to form events and coincidences of events described by punctual (point-to-point) correlations in a four-dimensional manifold (M, g).

Because of the equivalence principle, these observables must be generated within a volume of finite spatial extent from a given spatial length L and propagated—through a chain of events—in the form of four-volumetric densities (or in geometrical sub-varieties) in the four-dimensional manifold (M, g) to a remotely located finite region of space–time of likewise finite spatial/temporal extent—the observation (hyper-)volume $V \subseteq M$ over which they are volume integrated into observables, allowing the information carried by them to be extracted and decoded. More specifically, we will use the local split of $3 + 1$ in space and time where we will consider the integration of the field properties over a three-dimensional volume $V = L^3$. The volumetric density of every gravitational observable carried by the gravitational field is a linear combination of quantities that are second order (quadratic/bilinear) in the metric of the field, and/or of the derivatives of the field. To obtain these quantities one must integrate the density of a conserved quantity e.g., over a given 3D space-like hypersurface σ or over a four-dimensional interval, characterized by a given finite length L; there, the field equations are integrated and averaged to obtain the field observables we need to analyze the properties of space–time down to the Planck scales.

Let us consider a Lorentzian manifold as example, with cosmological constant Λ and then introduce a characteristic length L. This quantity is the proper length associated to the generic coordinate variation Δx written in terms of the metric tensor g, viz., $L \sim g^{1/2} \Delta x$. The Riemann tensor is then written in terms of the metric variations Δg, the covariant metric tensor g and the contrarvariant one, g^{-1}

$$R_{(4)}(g,L) \sim \frac{g^2}{L^2}\left(\Delta\left(\Delta g(g)^{-1}\right) + \left(\Delta g(g)^{-1}\right)^2\right) \qquad (3)$$

where the term $(\Delta g(g)^{-1})^2_{\{kl,im-km,il\}} = \Gamma^n_{kl}\Gamma^p_{im} - \Gamma^n_{km}\Gamma^p_{il}$ represents the affine connection and $\Delta(g^{-1}\Delta g)$ is the second derivatives of the metric tensor with respect to the coordinates, being $\partial^2_{kl} g = \frac{g}{L^2}\Delta_k \Delta_l g = \frac{g^2}{L^2}\Delta_k\left(\Delta_l g(g)^{-1}\right)$. The Ricci tensor and scalar are $R_{(2)} \sim g^{-1}R_{(4)}$ and $R \sim g^{-2}R_{(4)}$, the Einstein tensor is $G = R_{(4)}(g)^{-1}$ and Einstein's equations are $G + g\Lambda = T$, where T is the energy–momentum tensor.

By introducing a characteristic length L, if Einstein's equations hold down to the Planck scales, from the basic formulation of the Riemann tensor and Einstein equations we find that the field equations, integrated and averaged over a 3D space-like hypersurface σ with unit normal vector $n \sim g^{-1/2}$, obey an indetermination relationship that recalls Heisenberg's. Instead of focusing on the more general energy–tensor quantity (or the momentum vector), we consider for the sake of simplicity the scalar proper energy E, averaged over a proper volume L^3, which is given by the integral of the energy momentum tensor over a given proper volume element of a space-like 3D–hypersurface. This leads to the following formulation of the proper energy averaged over the given proper volume

$$\langle E - g\Lambda\rangle = \bar{E} \sim \frac{g^2}{L}R_{(4)} = L\left(\Delta\left(\Delta g(g)^{-1}\right) + \left(\Delta g(g)^{-1}\right)^2\right) \qquad (4)$$

If we rescale this relationship down to the Planck scale L_p, by defining the light crossing time as $\tau = L$ and the Planck Time τ_p, the Einstein equations retain their validity down to the Planck scale, even if metric fluctuations over a scale larger than L_p can occur. We find that these fluctuations can give rise to a relationship

$$\left(\frac{\tau_p}{\tau}\right)^2\left(\frac{\bar{E}\times\tau}{\hbar}\right) = \left(\frac{L_p}{L}\right)^2\left(\frac{\bar{E}\times\tau}{\hbar}\right) = \frac{L^2}{g^2}R_{(4)}(g,L) \qquad (5)$$

that holds down to the Planck scales. Fixing a characteristic spatial scale (or time), the relationship in Equation (5) corresponds to the introduction of fluctuations of the averaged quantity over L^3 of the proper energy $\bar{E}$. If we set $\bar{E} = \Delta E^*$ and $\tau = \Delta t$, we can write Equation (5) in a more familiar Heisenberg relationship that involves the Riemann tensor and the contribution from the dark energy

$$\Delta E^* \times \Delta t = \hbar\left(\frac{\tau}{\tau_p}\right)^2\frac{L^2}{g^2}R_{(4)}(g,L) = \frac{\hbar}{g^2}\left(\frac{L^2}{L_p}\right)^2 R_{(4)}(g,L) \qquad (6)$$

that at Planck scales becomes

$$\Delta E^* \times \Delta t = \hbar\frac{L_p^2}{g^2}R_{(4)}(g,L) = \hbar\left(\Delta\left(\Delta g(g)^{-1}\right) + \left(\Delta g(g)^{-1}\right)^2\right) \qquad (7)$$

where $\Delta E^* = \Delta E + \Delta g\,\Lambda + g\Delta\Lambda$ averaged on the volume L^3 of the 3D space-like hypersurface σ.

2.2. The Energy of the Gravitational Field

Dark energy and other different vacua are parameterized by the cosmological constant Λ. When $\Lambda > 1$, the equations describe an AdS space–time. The gravitational fluctuations are mainly expressed by the affine connection term $(\Delta g(g)^{-1})^2$ for any space–time. To describe the energy of the gravitational field, which is not defined as a global and conserved quantity, one has to introduce pseudotensorial quantities describing the energy trapped non-locally in the geometry. One example is the non-symmetric Einstein pseudotensor, which is constructed exclusively from the metric tensor and its first derivatives but is not suitable for our purposes. Instead, the Landau–Lifshitz pseudotensor

t_{ik} [62] permits us to write for the integrated non-local gravitational energy E_g that includes the contribution of the cosmological constant in terms of the curvature tensor. This quantity is quadratic in the connection and, for a general covariant component of the pseudotensor, averaged on the volume $V = L^3$ one obtains

$$\left\langle g^{-1}(t + \Lambda g) \right\rangle_{V=L^3} \sim \frac{E_g + E_\Lambda}{L^3} \tag{8}$$

where $E_\Lambda = g^{-1} \Lambda g L^3$ is the energy associated to the value of the cosmological constant and to dark energy. Considering that

$$|g| \, g^{-2}(t + \Lambda g) \sim \left(\frac{\Delta g}{\Delta x}\right)^2, \tag{9}$$

this relationship leads to a background curvature with fluctuations having wavelength $\lambda = L$ that can be interpreted as connections between events due to an exchange of virtual gravitons with wavelength λ and energy $\hbar/\lambda$ or, in the ER = EPR scenario, to the connection through an ER wormhole,

$$\left\langle \left(\Delta g(g)^{-1}\right)^2 \right\rangle_{V=L^3} \sim \frac{E_g + E_\Lambda}{L} = \left(\frac{\tau_p}{\tau}\right) \frac{E_g + E_\Lambda}{E_p} \tag{10}$$

where E_p is Planck's energy. In the ER = EPR hypothesis, these energy fluctuations would be considered as equivalent to the connection between two entangled events separated by the distance L, giving a paradoxical meaning to the exchange of a virtual graviton in terms of entanglement connections between events like in an emergent gravity scenario.

Following the already cited works by De Witt and the classical QG interpretation found in the literature [1–5], this term would describe a virtual graviton exchange between two events within a space–time connection. On the other hand, this term—that can be also interpreted in terms of a wormhole connection between the two events—with the exchange of at least 1 qbit of information (in the ER = EPR conjecture) would correspond to the entanglement of two particles. If ER wormholes are equivalent to a monogamous connection between the two events [20], as realized through a virtual graviton exchange, one could state that entanglement of EPR states can derive from the exchanges of virtual gravitons between two events. From another perspective, entangled states should be provided by a mixed state between the two entangled pairs with that of the virtual graviton. The question is what is the correct perspective?

From Einstein's equations, the observable averaged total energy of a metric fluctuation over the volume V on a scale L, becomes

$$E \sim \left(\frac{\tau}{\tau_p}\right) \left\langle \Delta \left(\Delta g(g)^{-1}\right) \right\rangle_V E_p + E_g + E_\Lambda \tag{11}$$

and is made with the energy of geometry and vacuum and energy of interaction expressed in terms of gradients of the geometry fluctuations, second order derivatives of the metric tensor, as in the Riemann tensor that make the connection between observers.

2.3. Planck-Scale Wormhole Connections

In a local neighborhood of a given event $\{x^i\}_0$, one performs a discrete infinite denumerable 3 + 1 local slicing of the space–time with time steps a Planck time unit. To the initial event $\{x^i\}_0$ corresponds the slice $N = 0$. The N-th slice corresponds to the time $\tau_N = (N+1)\tau_p$, building up a symbolic dynamics of space–time events. The energy of the gravitational perturbation in the N-th slice is

$$(E_g + E_\Lambda)_N \sim \frac{E_p}{N+1} \tag{12}$$

the total energy is instead

$$E_N \sim (N+1) \left\langle \Delta \left(\Delta g(g)^{-1} \right) \right\rangle_{V,N} E_p + \frac{E_p}{N+1} \quad (13)$$

for $N \to 0$ the energy fluctuation becomes $E_N \to E_p$ for which, by definition, $E_N \tau_N \sim \hbar$ and the metric tidal fluctuations tend to zero—space–time at Planck lengths is homogeneous and isotropic and the local geometry depends only on the energy of fluctuations in space–time and from the energy of the cosmological constant. This because $\Delta g/g \to 1$ are both on the order of the Planck scale. This is the reason why the field does not diverge and no singularities are present.

For $N \to \infty$ the dominant energy is that of tidal fluctuations at scales larger than L_p accompanied with that of the cosmological constant when integrated over the metric and remains as a constant function over the volume of integration; E_g becomes instead negligible. This means that for a process connecting two events lasting a time τ, the amount of energy does not entirely contribute to the vacuum energy but it is partially spent in geometry in this process, involving the dark energy contribution expressed by the cosmological constant Λ. Recalling Heisenberg principle from Equation (5), the larger is the energy fluctuation, the smaller results the space/time interval fluctuation.

In the neighborhood of Planck scales, when $N > 0$, the curvature of space–time remains finite and the Riemann tensor can be written as

$$R_{(4)}(g,L) \sim \frac{E_p}{\hbar} \left(\frac{\tau_p}{\tau} \right)^2 \frac{g^2}{L^2} \quad (14)$$

and the Ricci scalar is

$$R(g,L) \sim \frac{1}{L^2} \frac{E_p}{\hbar} \left(\frac{\tau_p}{\tau} \right)^2 \quad (15)$$

that for $L \to L_p$ we have $R(g,L) \to 1/L_p^2$ with the result that at Planck scales there is no singularity in the curvature and the gravitational radius becomes

$$R_g = 2 \frac{E \tau L_p^2}{\hbar L} \quad (16)$$

that is written as $R_g = 2L_p$, which corresponds to elementary wormhole connections at the Planck scale and finding a trivial equivalence with the corresponding Penrose diagrams. Directly from Einstein's equations we find that, at Planck scales, the singularities expected from quantum gravity can be interpreted in terms of wormhole connections between the events, as required in the ER = EPR conjecture and obtain an indetermination relationship shown in Equation (6) involving the Riemann tensor and geometry fluctuations. In this view, wormhole and equivalent EPR connections can also be formally equivalent to an exchange of a virtual graviton at scales larger than Planck scale, whilst any group of superimposed states below Planck scales, instead, will be indistinguishable and therefore entangled.

2.4. Tests for the ER = EPR Conjecture

If we suppose the validity of the ER = EPR conjecture, the geometry fluctuations present at Planck scales may be revealed with quantum entanglement. By applying the indetermination relationship that involves the Riemann tensor expressed in Equations (6) and (7), we argue that one can obtain information about the fluctuations of space–time and determine whether a characteristic scale like the Planck's one is present, as expected in QG.

If space–time is discrete, its discreteness is expected to be characterized by a typical scale of space and/or time: There exist a minimum time interval t_p and a minimum length L_p where wormholes connections—equivalent to entangled states between two or more regions of space–time—connect different events or space–times. If events/space–times are connected with intervals smaller than L_p and t_p, they would be entangled and actually be the same event or the same space–time. Their quantum superposition can exist and collapse after a finite time interval and the properties of wormhole connections are reflected in the properties of the corresponding EPR states also when they connect events at scales larger than the Planck scale. This scenario is different from Penrose's assumptions [55], where space–time is thought to be continuous and the quantum superposition of space–times result unfeasible leading to the gravitational collapse.

Noe we propose to test the ER = EPR scenario by using the Heisenberg uncertainty principle applied to pairs (or groups) of entangled particles and including the additional indetermination introduced by quantum gravity effects. The generalized Heisenberg's uncertainty principle for the momentum p and the position x that includes the existence of a characteristic length L_p such as the Planck scale, or any other scale typical that can be found in certain quantum gravity models, is given by [63,64]

$$\Delta x = \Delta x_{QM} + \Delta x_{GR} \geq \frac{\hbar}{2\Delta p} + k\,\Delta p \tag{17}$$

the existence of a minimum interval in space–time is revealed by a deviation from the classical term due to quantum mechanics only, Δx_{QM}. The quantum gravity term, Δ_{GR}, due to the existence of a characteristic length L_p and to the properties of the gravitational field, can be characterized instead by a parameter k, a constant characteristic of the quantum theory of gravitation here considered. To give an example, in a string theory scenario, $k = \alpha Y$, where α is the string tension and Y a constant that depends on the theory. In our case, following [65–67], one can find that $k = 2L_p^2/\hbar$. From our calculations that involve the Riemann tensor, we find that $\Delta x_{QG} = 2E_p/\Delta E^*$ and thus $\Delta p = \hbar E_p/\Delta E^* L_p^2$, a term that includes the effects of dark energy in the term ΔE^* too.

We write now the Heisenberg relationship for sets of N–particle entangled states. Following [68–72], consider first a couple of entangled particles with positions x_1 and x_2 and momenta p_1 and p_2, respectively. For $N = 2$, the classical indetermination principle is

$$\Delta(x_1, x_2)_{QM}^2 = \left[\Delta(x_1)^2 + \Delta(x_2)^2\right] \times \left[\Delta(p_1)^2 + \Delta(p_2)^2\right] \geq \frac{\hbar^2}{4}. \tag{18}$$

In the simplest case, where $\Delta(x_1) = \Delta(x_2) = \Delta x_e$ and $\Delta(p_1) = \Delta(p_2) = \Delta p_e$, the uncertainty relationship becomes $(\Delta x_e)^2 (\Delta p_e)^2 \geq \hbar^2$. For N identical entangled states, the extended indetermination principle becomes

$$(\Delta x_e)^2 (\Delta p_e)^2 \geq \frac{N^2 \hbar^2}{4} \tag{19}$$

and when we include the effects of the gravitational field one obtains

$$\Delta x = \Delta(x_1, x_2)_{QM} + \Delta(x_1, x_2)_{GR} \geq \frac{N\hbar}{2\Delta p} + \frac{2NL_p^2 \Delta p}{\hbar}. \tag{20}$$

By assuming that Einstein's equations retain their validity down to the Planck scales and that wormhole connections represent the building blocks of the physics of the gravitational field at and below Planck scales (t_p and L_p), ER = EPR links connecting any space–time (or event) with a difference smaller than t_p and L_p mean that different space–times and events are physically identical, and then in principle undetectable and entangled. Instead, in a region with radius R, the spatial difference of two space–times/events is $\Delta L = 2L_p^2 \Delta E^*/\hbar c$, and the difference of their corresponding space–times is the difference of the proper spatial sizes of the regions occupied by them and the time of the wavefunction collapse is on the order of $\tau_c \sim 2\hbar E_p/(\Delta E^*)^2$.

If the properties of ER = EPR links remain valid from the Planck up to the macroscopic scales, where entanglement can be observed in the lab, the term Δx_{QG} in the Heisenberg relationship expressed in Equation (20) is expected to reveal the properties of the wormhole structure of space–time from a deep analysis of the wavefunction collapse of an entangled pair. In other words, the deviation from the quantity Δx_{QM} of the classical Heisenberg principle would reveal the fuzziness space–time or, better, of the point-by-point identification of the spatial section of the two events/space–times, better evident with a set of a large number of N entangled quanta like a Schrödinger cat.

From the point of view of relativistic quantum information discipline, entanglement and wormholes are expected to create space–time and entanglement events (and space–times) through quantum information—information that emerges from the connection of quantum bits. In fact, from a quantum-computational interpretation of space–time entanglement, in a foliation of space–time, the quantum fluctuations of the metric present on the slice n can be interpreted as wormhole connections between one Planckian pixel in the slice n with that one present in the $n-1$ slice. Following [73], the holographic principle suggests that such a geometrical connection is space–time entanglement. If not entangled, following Penrose's argumentation only the quantum superposition of two space–times with a difference larger than the minimum sizes can not exist, and should collapse instantaneously. If they are connected by an ER wormhole they should obey the indetermination relationship expressed in Equation (20).

To verify possible additional anomalies in the indetermination principle introduced by the ER = EPR conjecture one may instead want to consider to measure the time/energy entangled states and study the time of collapse as a function of their energy differences. This may explain why the wavefunction collapse of an EPR pair is not always instantaneous, as it may depend on the geometry fluctuations. Moreover, one has to also consider the effects introduced by the presence of the cosmological constant, of the information encoded and shared between the entangled quantum states and their relationship with the gravitational information entropy that go beyond the purpose of the present work.

Of course an experimenter has to consider that EPR states depend on the choice of reference frames and that Bell's inequalities are preserved in certain reference frames only, and should also consider the effects of simultaneity and include in the experiment the additional macroscopic effects induced by the gravitational field at large scales in the presence of massive bodies. As an example, simultaneity is responsible for the uncertainty of the ordering of non-local wavefunction collapse when the relativistic effects cannot be neglected. In any case, if a time measurement performed with an entangled pair of photons is seen as simultaneous in one shared reference frame, then the result of this measure can be considered simultaneous to all measuring observers who do not share a reference frame. The inversion of the temporal order due to simultaneity is impossible to determine, the attempt to measure this effect will unavoidably introduce an uncertainty in the result. There is no need to have any preferred reference frame for the wavefunction collapse of entangled states. If an experimenter tries to determine the exact reference frame where the wavefunction collapsed, the measurement process will unavoidably introduce an uncertainty that would make impossible the identification of the "exact" reference frame. Obviously, if one can determine the order of the measurement in a shared reference frame it can result like that in certain reference frames and, instead, indeterminate in the other reference frames [74,75].

3. Discussion and Conclusions

It is one of the great merits of Albert Einstein to have investigated the possibility of a multiple-connected space–time and in theoretical physics there is a long tradition of studying quantum behavior in spaces of this type [76]. These lines of research have progressively merged into the quantum study of wormholes, assuming a decisive relevance not only for the study of the structure of the GR and its cosmological implications, but has given the question a decisive configuration as regards the relations of "coexistence" peaceful between QM and GR. In this work we proposed a formal technique for the study of the quantum effects of a wormhole within the conjecture ER = EPR. We then considered different scenarios from the original Susskind and Maldacena one, in particular those related to the dS space, which seems to be a much more promising ground for the study of the emergence of classical information starting from a quantum background where time is not defined [56,77–79].

These reflections suggest that an effective generalization of the physical meaning of ER = EPR requires a different and more complex philosophy on the emergence of physical space–time as a holographic "settlement" of temperature/energy scales, and the use of well-known techniques in QFT [80,81].

In other words, these scenarios suggest that the idea of transition of the metric suggested by Sacharov may be the most "natural" way to characterize non-locality in a metric formalism. The assumption of ER = EPR would be only one of the aspects of a more general phenomenon of Raum–Zeit–Materie production starting from a non-local Euclidean background through quantum computation procedures. The observable part of space time would therefore, in a rather literal sense, result in a thin layer of ice emerging from an ocean of non-locality and the extension of ER = EPR conjecture to Euclidean non-locality may extend its domain from the original AdS/CFT scenario.

Finally, we suggest readers consider the conjecture ER = EPR within the scenario of de Sitter's projective cosmology, described by Hartle-Hawking boundary conditions as Nucleation by Sitter Vacuum [57]. In this cosmological approach one can define the localization conditions in time of the particles starting from an Euclidean pre-space that models a non-local phase. Using the Bekenstein relation, it is possible to identify the area of the micro-horizon $A = (c\theta_0)^2 \simeq 10^{-26}$ cm^2, where theta is the chronon, chosen as time scale of the baryonic location. In this case the construction of wormholes applies to a scale much larger than the Planck length. In this case the wormholes are defined by a transition of the metric similar to that hypothesized in the classical work by Sacharov in 1984 [82].

Anyway, the wormhole structure of space–time could in principle be characterized by the extended Heisenberg principle through a deep study of the wavefunction collapse of entangled particles and reveal possible scenarios of QG and cosmology or emergent gravity theories where the exchange of a virtual graviton could also be interpreted in terms of entanglement. At Planck scales wormhole connections would avoid the gravitational collapse and singularities. Moreover, the exchange of a virtual graviton would become equivalent to a wormhole connection and/or entanglement between two or more events. From this we can argue that the ER = EPR conjecture alone, as it is, cannot fully explain without experimental results whether Planck-scale phenomenology can be revealed through entanglement or that gravity and space–time are emergent physical quantities.

Author Contributions: Writing—original draft, F.T. and I.L. All authors have read and agreed to the published version of the manuscript.

Acknowledgments: We dedicate this work to Dino and Gep. F.T. acknowledges ZKM and Peter Weibel for the financial support.

Conflicts of Interest: The authors declare no conflict of interest.

References

1. DeWitt, B.S. Quantum Theory of Gravity. I. The Canonical Theory. *Phys. Rev.* **1967**, *160*, 1113. [CrossRef]
2. DeWitt, B.S. Quantum Theory of Gravity. II. The Manifestly Covariant Theory. *Phys. Rev.* **1967**, *162*, 1195. [CrossRef]
3. DeWitt, B.S. Quantum Theory of Gravity. III. Applications of the Covariant Theory. *Phys. Rev.* **1967**, *162*, 1239. [CrossRef]
4. DeWitt, B.S.; DeWitt, C.M. The Quantum Theory of Interacting Gravitational and Spinor Fields. *Phys. Rev.* **1952**, *87*, 116. [CrossRef]
5. DeWitt, B.S. Approximate Effective Action for Quantum Gravity. *Phys. Rev. Lett.* **1981**, *47*, 1647. [CrossRef]
6. Rosenfeld, L. Zur quantelung der wellenfelder. *Ann. Phys.* **1930**, *5*, 113. [CrossRef]
7. Rosenfeld, L.Z. Über die Gravitationswirkungen des Lichtes. *Physik* **1930**, *65*, 589. [CrossRef]
8. Tanabashi, M.; Hagiwara, K.; Hikasa, K.; Nakamura, K.; Sumino, Y.; Takahashi, F.; Tanaka, J.; Agashe, K.; Aielli, G.; Amsler, C.; et al. Review of Particle Physics. *Phys. Rev. D* **2018**, *98*, 030001. [CrossRef]
9. Rovelli, C. Loop Quantum Gravity. *Living Rev. Relativ.* **2008**, *11*, 5. [CrossRef] [PubMed]
10. Dzhunushaliev, V.D. Multidimensional Ggeometrical Model of the Renormalized Electrical Charge with Splitting off the Extra Coordinates. *Mod. Phys. Lett. A* **1998**, *13*, 2179. [CrossRef]

11. Dzhunushaliev, V.D.; Singleton, D. Wormholes and Flux Tubes in 5D Kaluza-Klein Theory. *Phys. Rev. D* **1999**, *59*, 064018. [CrossRef]
12. Maldacena, J.; Susskind, L. Cool horizons for entangled black holes. *Fortschr. Phys.* **2013**, *61*, 781–811. [CrossRef]
13. Einstein, A.; Rosen, N. The Particle Problem in the General Theory of Relativity. *Phys. Rev.* **1935**, *48*, 73–77. [CrossRef]
14. Einstein, A.; Podolsky, B.; Rosen, N. Can Quantum-Mechanical Description of Physical Reality Be Considered Complete? *Phys. Rev.* **1935**, *47*, 777–780. [CrossRef]
15. Susskind, L. Copenhagen vs Everett, Teleportation, and ER = EPR. *Fortschr. Phys.* **2016**, *64*, 551–564. [CrossRef]
16. Maldacena, J. The large N limit of superconformal field theories and supergravity. *Adv. Theor. Math. Phys.* **1998**, *2*, 231. [CrossRef]
17. Aharony, O.; Gubser, S.S.; Maldacena, J.; Ooguri, H.; Oz, Y. Large N Field Theories, String Theory and Gravity. *Phys. Rep.* **2000**, *323*, 183–386. [CrossRef]
18. Biquard, O. *AdS/CFT Correspondence: Einstein Metrics and Their Conformal Boundaries*. EMS IRMA Lectures in Mathematics and Theoretical Physics; European Mathematical Society: Strasbourg, France, 2005; ISBN 978-3-03719-013-5.
19. Witten, E. Anti de Sitter space and holography. *Adv. Theor. Math. Phys.* **1998**, *2*, 253–291. [CrossRef]
20. Gharibyan, H.; Penna, R.F. Are entangled particles connected by wormholes? Support for the ER = EPR conjecture from entropy inequalities. *Phys. Rev. D* **2014**, *89*, 066001. [CrossRef]
21. van Raamsdonk, M. Building up space–time with quantum entanglement. *Gen. Relativ. Gravit.* **2010**, *42*, 2323–2329. [CrossRef]
22. Verlinde, E.P. On the Origin of Gravity and the Laws of Newton. *JHEP* **2011**, 29. [CrossRef]
23. Cao, C.; Carroll, S.M. Bulk entanglement gravity without a boundary: Towards finding Einstein's equation in Hilbert space. *Phys. Rev. D* **2018**, *97*, 086003. [CrossRef]
24. Bell, J.S. On the Einstein-Poldolsky-Rosen paradox. *Physics* **1964**, *1*, 195–200. [CrossRef]
25. Gottfried, K. *John S. Bell on the Foundations of Quantum Mechanics*; Bell, M., Gottfried, K., Veltman, M., Eds.; World Scientific Publishing Company: Singapore, 2001.
26. Horodecki, R.; Horodecki, P.; Horodecki, M.; Horodecki, K. Quantum entanglement. *Rev. Mod. Phys.* **2009**, *81*, 865. [CrossRef]
27. Bouwmeester, D.; Ekert, A.; Zeilinger, A. (Eds.) *The Physics of Quantum Information*; Springer: Berlin/Heidelberg, Germany, 2000; ISBN 978-3-642-08607-6.
28. Bertlmann, R.; Zeilinger, A. (Eds.) *Quantum [Un]Speakables, from Bell to Quantum Information*; Springer: Berlin/Heidelberg, Germany, 2002; ISBN 978-3-662-05032-3.
29. Bertlmann, R.; Zeilinger, A. (Eds.) *Quantum [Un]Speakables II*; Springer Nature AG: Basel, Switzerland, 2017; ISBN 978-3-319-38987-5.
30. Visser, M. *Lorentzian Wormholes: From Einstein to Hawking*; AIP: Woodbury, NY, USA, 1995; 412p.
31. Maldacena, J.; Milekhin, A.; Popov, F. Traversable wormholes in four dimensions. *arXiv* **2018**, arXiv:1807.04726.
32. Maldacena, J.; Qi, X.-L. Eternal traversable wormhole. *arXiv* **2018**, arXiv:1804.00491.
33. Horowitz, G.T.; Marolf, D.; Santos, J.E.; Wang, D. Creating a Traversable Wormhole. *arXiv* **2019**, arXiv:1904.02187.
34. Bao, N.; Chatwin-Davies, A.; Pollack, J.; Remmen, G.N. Traversable wormholes as quantum channels: Exploring CFT entanglement structure and channel capacity in holography. *JHEP* **2018**, 71. [CrossRef]
35. Bueno, P.; Cano, P.A.; Goelen, F.; Hertog, T.; Vercnocke, B. Echoes of Kerr-like wormholes. *Phys. Rev. D* **2018**, *97*, 024040. [CrossRef]
36. Cardoso, V.; Franzin, E.; Pani, P. Is the gravitational-wave ringdown a probe of the event horizon? *Phys. Rev. Lett.* **2016**, *116*, 171101; Erratum *Phys. Rev. Lett.* **2016**, *117*, 089902. [CrossRef]
37. Konoplya, R. How to tell the shape of a wormhole by its quasinormal modes. *Phys. Lett. B* **2018**, *784*, 43–49. [CrossRef]
38. Bekenstein, J.D. Black holes and entropy. *Phys. Rev. D* **1973**, *7*, 2333–2346. [CrossRef]
39. Hawking, S.W. Particle creation by black holes. *Commun. Math. Phys.* **1975**, *43*, 199–220. [CrossRef]

40. Hubeny, V.E.; Rangamani, M.; Takayanagi, T. A Covariant Holographic Entanglement Entropy Proposal. *JHEP* **2007**. [CrossRef]
41. Sethi, S.S. Notes at URL. Available online: http://theory.uchicago.edu/~sethi/Teaching/P483-W2018/Intro%20to%20the%20RT.pdf (accessed on 18 December 2019).
42. Boyle, L.; Finn, K.; Turok, N. CPT-Symmetric Universe. *Phys. Rev. Lett.* **2019**, *121*, 251301. [CrossRef]
43. Narain, G.; Zhang, H.Q. Non-locality effect on the entanglement entropy in deSitter. *JCAP* **2019**, *06*, 012. [CrossRef]
44. Chena, P.; Hua, Y.-C.; Yeoma, D.-H. Fuzzy Euclidean wormholes in de Sitter space. *JCAP* **2017**, *1707*, 001. [CrossRef]
45. Maldacena, J.; Pimentel G.L. Entanglement entropy in de Sitter space. *JHEP* **2013**, *1302*, 038. [CrossRef]
46. Narayan, K. de Sitter entropy as entanglement. *Int. J. Mod. Phys. D* **2019**. [CrossRef]
47. Arias, C.; Diaz, F.; Sundell, P. De Sitter Space and Entanglement. *arXiv* **2019**, arXiv:1901.04554.
48. Dong, X.; Silverstein, E.; Torroba, G. De Sitter holography and entanglement entropy. *JHEP* **2018**, 50. [CrossRef]
49. Dimopoulos, S.; Kachru, S.; Kaloper, N.; Lawrence, A.; Silverstein, E. Generating small numbers by tunneling in multithroat compactifications. *Int. J. Mod. Phys. A* **2004**, *19*, 2657–2704. [CrossRef]
50. Dimopoulos, S.; Kachru, S.; Kaloper, N.; Lawrence, A.E.; Silverstein, E. Small numbers from tunneling between brane throats. *Phys. Rev. D* **2001**, *64*, 121702. [CrossRef]
51. Polchinski, J.; Silverstein, E. Dual Purpose Landscaping Tools: Small Extra Dimensions in AdS/CFT. In *Strings, Gauge Fields, and the Geometry Behind. The Legacy of Maximilian Kreuzer*; Rebhan, A., Katzarkov, L., Knapp, J., Rashkov, R., Scheidegger, E., Eds.; World Scientific: Singapore, 2011; ISBN-13: 978-9814412544.
52. Vasiliev, M. Higher-Spin Gauge Theories in Four, Three and Two Dimensions. *Int. J. Mod. Phys. D* **1996**, *5*, 763–797. [CrossRef]
53. Anninos, D.; Hartman, T.; Strominger, A. Higher Spin Realization of the dS/CFT Correspondence. *CQG* **2016**, *34*, 015009. [CrossRef]
54. Dong, X.; Zhou, L. Spacetime as the optimal generative network of quantum states: A roadmap to QM=GR? *arXiv* **2018**, arXiv:1804.07908.
55. Penrose, R. On gravity's role in quantum state reduction. *Gen. Relativ. Gravit.* **1996**, *28*, 581–600. [CrossRef]
56. Licata, I.; Chiatti, L. Event-Based Quantum Mechanics: A Context for the Emergence of Classical Information. *Symmetry* **2019**, *11*, 181. [CrossRef]
57. Feleppa, F.; Licata, I.; Corda, C. Hartle-Hawking boundary conditions as Nucleation by de Sitter Vacuum. *Phys. Dark Universe* **2019**, *26*, 100381. [CrossRef]
58. Strominger, A. The dS/CFT Correspondence. *arXiv* **2001**, arXiv:hep-th/0106113v2.
59. Schwinger, J.; Deraad, L.L., Jr.; Milton, K.A.; Tsaiyang, W. *Classical Electrodynamics*; Perseus Books: Reading, MA, USA, 1965; ISBN 0-7382-0056-5.
60. Thidé, B. *Electromagnetic Field Theory*, 2nd ed.; Dover Publications, Inc.: Mineola, NY, USA, 2011.
61. Tamburini, F.; De Laurentis, M.; Licata, I. Radiation from charged particles due to explicit symmetry breaking in a gravitational field. *Int. J. Geom. Meth. Mod. Phys.* **2018**, *15*, 1850122. [CrossRef]
62. Landau, L.D.; Lifshitz, E.M. *The Classical Theory of Fields*, 4th ed.; Butterworth-Heinemann: Oxford, UK, 1975; Volume 2, ISBN 978-0-7506-2768-9.
63. Kempf, A.; Mangano, G.; Mann, R.B. Hilbert space representation of the minimal length uncertainty relation. *Phys. Rev. D* **1995**, *52*, 1108. [CrossRef] [PubMed]
64. Scardigli, F. Generalized Uncertainty Principle in Quantum Gravity from Micro-Black Hole Gedanken Experiment. *Phys. Lett. B* **1999**, *452*, 39–44. [CrossRef]
65. Adler, R.J.; Santiago, D.I. On Gravity and the Uncertainty Principle. *Mod. Phys. Lett. A* **1999**, *14*, 1371–1381. [CrossRef]
66. Garay, L.J. Quantum gravity and minimum length. *Int. J. Mod. Phys. A* **1995**, *10*, 145. [CrossRef]
67. Shan, G. A Model of Wavefunction Collapse in Discrete Space-Time. *IJTP* **2006**, *45*, 10. [CrossRef]
68. Blado, G.; Herrera, F.; Erwin, J. Quantum Entanglement and the Generalized Uncertainty Principle. *arXiv* **2017**, arXiv:1706.10013.
69. Zeng, J.; Lei, Y.; Pei, S.Y.; Zeng, X.C. CSCO Criterion for Entanglement and Heisenberg Uncertainty Principle. *arXiv* **2013**, arXiv:1306.3325.

70. Rigolin, G. Uncertainty relations for entangled states. *Found. Phys. Lett.* **2002**, *15*, 293.:1021039822206. [CrossRef]
71. Rigolin, G. Entanglement, Identical Particles and the Uncertainty Principle. *Commun. Theor. Phys.* **2016**, *66*, 201. [CrossRef]
72. Prevedel, R.; Hamel, D.R.; Colbeck, R.; Fisher, K.; Resch, K.J. Experimental investigation of the uncertainty principle in the presence of quantum memory and its application to witnessing entanglement. *Nat. Phys.* **2011**, *7*, 757–761. [CrossRef]
73. Zizzi, P. Entangled Space-Time. *Mod. Phys. Lett. A* **2018**, *33*, 1850168. [CrossRef]
74. Resconi, G.; Licata, I.; Fiscaletti, D. Unification of Quantum and Gravity by Non Classical Information Entropy Space. *Entropy* **2013**, *15*, 3602–3619. [CrossRef]
75. Olson, S.J.; Dowling, J.P. Probability, unitarity, and realism in generally covariant quantum information. *arXiv* **2007**, arXiv:0708.3535.
76. Ho, V.B.; Morgan, M.J. Mathematical and General Quantum mechanics in multiply–connected spaces. *J. Phys. A* **1996** *29*, 7.
77. Vistarini, T. Holographic space and time: Emergent in what sense? *Stud. Hist. Philos. Sci. Part B Stud. Hist. Philos. Mod. Phys.* **2017**, *59*, 126–135. [CrossRef]
78. Rovelli, C. Quantum mechanics without time: A model. *Phys. Rev. D* **1990** *42*, 2638. [CrossRef]
79. Qi, X.-L. Exact holographic mapping and emergent space–time geometry. *arXiv* **2013**, arXiv:1309.6282v1.
80. Barvinsky, A.O. Aspects of nonlocality in quantum field theory, quantum gravity and cosmology. *Mod. Phys. Lett. A* **2015**, *30*, 1540003 . [CrossRef]
81. Garrett Lisi, A. Quantum mechanics from a universal action reservoir. *arXiv* **2006**, arXiv:physics/0605068.
82. Sacharov, A. Cosmological Transitions with a Change in Metric Signature. *Usp. Fiz. Nauk* **1991**, *161*, 94–104. [CrossRef]

© 2019 by the authors. Licensee MDPI, Basel, Switzerland. This article is an open access article distributed under the terms and conditions of the Creative Commons Attribution (CC BY) license (http://creativecommons.org/licenses/by/4.0/).

Article

The Information Loss Problem: An Analogue Gravity Perspective

Stefano Liberati [1,2,3,*], Giovanni Tricella [1,2,3,*] and Andrea Trombettoni [1,2,3,4,*]

1   SISSA-International School for Advanced Studies,Via Bonomea 265, 34136 Trieste, Italy
2   INFN Sezione di Trieste, Via Valerio 2, 34127 Trieste, Italy
3   IFPU-Institute for Fundamental Physics of the Universe, Via Beirut 2, 34014 Trieste, Italy
4   CNR-IOM DEMOCRITOS Simulation Center, Via Bonomea 265, I-34136 Trieste, Italy
*   Correspondence: liberati@sissa.it (S.L.); gtricell@sissa.it (G.T.); andreatr@sissa.it (A.T.)

Received: 21 August 2019; Accepted: 22 September 2019; Published: 25 September 2019

Abstract: Analogue gravity can be used to reproduce the phenomenology of quantum field theory in curved spacetime and in particular phenomena such as cosmological particle creation and Hawking radiation. In black hole physics, taking into account the backreaction of such effects on the metric requires an extension to semiclassical gravity and leads to an apparent inconsistency in the theory: the black hole evaporation induces a breakdown of the unitary quantum evolution leading to the so-called information loss problem. Here, we show that analogue gravity can provide an interesting perspective on the resolution of this problem, albeit the backreaction in analogue systems is not described by semiclassical Einstein equations. In particular, by looking at the simpler problem of cosmological particle creation, we show, in the context of Bose–Einstein condensates analogue gravity, that the emerging analogue geometry and quasi-particles have correlations due to the quantum nature of the atomic degrees of freedom underlying the emergent spacetime. The quantum evolution is, of course, always unitary, but on the whole Hilbert space, which cannot be exactly factorized a posteriori in geometry and quasi-particle components. In analogy, in a black hole evaporation one should expect a continuous process creating correlations between the Hawking quanta and the microscopic quantum degrees of freedom of spacetime, implying that only a full quantum gravity treatment would be able to resolve the information loss problem by proving the unitary evolution on the full Hilbert space.

Keywords: analogue gravity; Bose-Einstein condensation; information loss; cosmological particle creation

1. Introduction

Albeit being discovered more than 40 years ago, Hawking radiation is still at the center of much work in theoretical physics due to its puzzling features and its prominent role in connecting general relativity, quantum field theory, and thermodynamics. Among the new themes stimulated by Hawking's discovery, two have emerged as most pressing: the so-called transplanckian problem and the information loss problem.

The transplanckian problem stems from the fact that infrared Hawking quanta observed at late times at infinity seems to require the extension of relativistic quantum field theories in curved spacetime well within the UV completion of the theory, i.e., the Hawking calculation seems to require a strong assumption about the structure of the theory at the Planck scale and beyond.

With this open issue in mind, in 1981, Unruh introduced the idea to simulate in condensed matter systems black holes spacetime and the dynamics of fields above them [1]. Such analogue models of gravity are provided by several condensed-matter/optical systems in which the excitations propagate in an effectively relativistic fashion on an emergent pseudo-Riemannian geometry induced by the medium. Indeed, analogue gravity has played a pivotal role in the past years by providing a test bench for many open issues in quantum field theory in curved spacetime and in demonstrating the robustness of Hawking radiation and cosmological primordial spectrum of perturbations stemming from inflation against possible UV completions of the theory (see, e.g., the work by the authors of [2] for a comprehensive review). In recent years, the same models have offered a valuable framework within which current ideas about the emergence of spacetime and its dynamics could be discussed via convenient toy models [3–6].

Among the various analogue systems, a preeminent role has been played by Bose–Einstein condensates (BEC), because these are macroscopic quantum systems whose phonons/quasi-particle excitations can be meaningfully treated quantum mechanically. Therefore, they can be used to fully simulate the above mentioned quantum phenomena [7–9] and also as an experimental test bench of these ideas [10–13].

In what follows, we shall argue that these systems cannot only reproduce Hawking radiation and address the transplanckian problem, but can also provide a precious insight into the information loss problem. For gravitational black holes, the latter seems to be a direct consequence of the backreaction of Hawking radiation, leading to the decrease of the black hole mass and of the region enclosed by the horizon. The natural endpoint of such process into a complete evaporation of the object leads to a thermal bath over a flat spacetime, which appears to be incompatible with a unitary evolution of the quantum fields from the initial state to the final one. (We are not considering here alternative solutions such as long-living remnants, as these are as well problematic in other ways [14–18], or they imply deviations from the black hole structure at macroscopic scale, see, e.g., the work by the authors of [19]).

Of course, the BEC system at the fundamental level cannot violate unitary evolution. However, it is obvious that one can conceive analogue black holes provided with singular regions for the emergent spacetime where the description of quasi-particles propagating on an analogue geometry fails. (For example, one can describe flows characterized by regions where the hydrodynamical approximation fails even without necessarily having loss of atoms from the systems). In such cases, despite the full dynamics being unitary, it seems that a trace over the quasi-particle falling in these "analogue singularities" would be necessary, so leading to an apparent loss of unitarity from the analogue system point of view. The scope of the present investigation is to describe how such unitarity evolution is preserved on the full Hilbert space.

However, in addressing the information loss problem in gravity, the spacetime geometry and the quantum fields are implicitly assumed to be separated sectors of the Hilbert space. In the BEC analogue, this assumption is reflected in the approximation that the quantum nature of the operator $\hat{a}_{k=0}$, creating particles in the background condensate, can be neglected. Therefore, in standard analogue gravity, a description of the quantum evolution on the full Hilbert space seems precluded.

Nonetheless, it is possible to retain the quantum nature of the condensate operator as well as to describe their possible correlations with quasi-particles within an improved Bogoliubov description, namely, the number-conserving approach [20]. Remarkably, we shall show that the analogue gravity framework can be extended also in this context. Using the simpler setting of a cosmological particle creation, we shall describe how this entails the continuous generation of correlations between the condensate atoms and the quasi-particles. Such correlations are responsible for (and in turn consequence of) the nonfactorizability of the Hilbert space and are assuring in any circumstances the unitary evolution of the full system. The lesson to be drawn is that in gravitational systems only a full quantum gravity description could account for the mixing between gravitational and matter quantum degrees of freedom and resolve, in this way, the apparent paradoxes posed by black hole evaporation in quantum field theory on curved spacetime.

The paper is organized as follows. In Section 2, we briefly recall the analogue gravity model for a nonrelativistic BEC in mean-field approximation. In Section 3, we review the time-dependent orbitals formalism and how it is employed in the general characterization of condensates and in the description of the dynamics. In Section 4, we introduce the number-conserving formalism. In Section 5, we discuss the conditions under which we can obtain an analogue gravity model in this general case. Finally, we analyze, in Section 6, how in analogue gravity the quasi-particle dynamics affects the condensate and, in Section 7, how the condensate and the excited part of the full state are entangled, showing that the unitarity of the evolution is a feature of the system considered in its entirety. Section 8 is devoted to a discussion of the obtained results and of the perspectives opened by our findings.

2. Analogue Gravity

In this Section, we briefly review how to realize a set-up for analogue gravity [2] with BEC in the Bogoliubov approximation, with a bosonic low-energy (nonrelativistic) atomic system. In particular, we consider the simplest case, where the interaction potential is given by a local 2-body interaction. The Hamiltonian operator, the equation of motion, and commutation relations in second quantization formalism are as usual:

$$H = \int dx \left[\phi^\dagger(x) \left(-\frac{\nabla^2}{2m} \phi(x) \right) + \frac{\lambda}{2} \phi^\dagger(x) \phi^\dagger(x) \phi(x) \phi(x) \right], \quad (1)$$

$$i\partial_t \phi(x) = [\phi(x), H] = -\frac{\nabla^2}{2m} \phi(x) + \lambda \phi^\dagger(x) \phi(x) \phi(x), \quad (2)$$

$$\left[\phi(x), \phi^\dagger(y) \right] = \delta(x - y). \quad (3)$$

For notational convenience, we dropped the time dependence of the bosonic field operator, ϕ, while retaining the dependence from the spatial coordinates x and y, and for simplicity, we omit the hat notation for the operators. Moreover, m is the atomic mass and the interaction strength λ, proportional to the scattering length [20], and could be taken as time-dependent. We also set $\hbar = 1$.

In the Bogoliubov approximation [20,21], the field operator is split into two contributions: a classical mean-field and a quantum fluctuation field (with vanishing expectation value),

$$\phi(x) = \langle \phi(x) \rangle + \delta\phi(x), \quad (4)$$

$$\left[\delta\phi(x), \delta\phi^\dagger(y) \right] = \delta(x - y). \quad (5)$$

The exact equations for the dynamics of these objects could be obtained from the full Equation (2), but two approximations ought to be considered:

$$i\partial_t \langle \phi \rangle = -\frac{\nabla^2}{2m} \langle \phi \rangle + \lambda \langle \phi^\dagger \phi \phi \rangle \approx -\frac{\nabla^2}{2m} \langle \phi \rangle + \lambda \overline{\langle \phi \rangle} \langle \phi \rangle \langle \phi \rangle, \quad (6)$$

$$i\partial_t \delta\phi = -\frac{\nabla^2}{2m} \delta\phi + \lambda \left(\phi^\dagger \phi \phi - \langle \phi^\dagger \phi \phi \rangle \right) \approx -\frac{\nabla^2}{2m} \langle \phi \rangle + 2\lambda \overline{\langle \phi \rangle} \langle \phi \rangle \delta\phi + \lambda \langle \phi \rangle^2 \delta\phi^\dagger \quad (7)$$

the bar denoting complex conjugation. The first equation is the Gross–Pitaevskii equation, and we refer to the second as the Bogoliubov–de Gennes (operator) equation. The mean-field term represents the condensate wave function, and the approximation in Equation (6) is to remove from the evolution the backreaction of the fluctuation on the condensate. The second approximation is to drop all the nonlinear terms (of order higher than $\delta\phi$) from Equation (7), which then should be diagonalized to solve the time evolution.

The standard set-up for analogue gravity is obtained by describing the mean-field of the condensate wave function, and the fluctuations on top of it, in terms of number density and phase, as defined in the so-called Madelung representation

$$\langle \phi \rangle = \rho_0^{1/2} e^{i\theta_0}, \quad (8)$$

$$\delta\phi = \rho_0^{1/2} e^{i\theta_0} \left(\frac{\rho_1}{2\rho_0} + i\theta_1 \right), \quad (9)$$

$$[\theta_1(x), \rho_1(y)] = -i\delta(x - y). \quad (10)$$

From the Gross–Pitaevskii equation, we obtain two equations for the real classical fields θ_0 and ρ_0:

$$\partial_t \rho_0 = -\frac{1}{m} \nabla \left(\rho_0 \nabla \theta_0 \right), \tag{11}$$

$$\partial_t \theta_0 = -\lambda \rho_0 + \frac{1}{2m} \rho_0^{-1/2} \left(\nabla^2 \rho_0^{1/2} \right) - \frac{1}{2m} \left(\nabla \theta_0 \right) \left(\nabla \theta_0 \right). \tag{12}$$

These are the quantum Euler equations for the superfluid. Equation (11) can be easily interpreted as a continuity equation for the density of the condensate, whereas Equation (12) is the Bernoulli equation for the phase of the superfluid, which generates the potential flow: the superfluid has velocity $\left(\nabla \theta_0 \right) / m$. From the Bogoliubov–de Gennes Equation, we obtain two equations for the real quantum fields θ_1 and ρ_1

$$\partial_t \rho_1 = -\frac{1}{m} \nabla \left(\rho_1 \nabla \theta_0 + \rho_0 \nabla \theta_1 \right), \tag{13}$$

$$\partial_t \theta_1 = -\left(\lambda \rho_0 + \frac{1}{4m} \nabla \left(\rho_0^{-1} \left(\nabla \rho_0 \right) \right) \right) \frac{\rho_1}{\rho_0} + \frac{1}{4m} \nabla \left(\rho_0^{-1} \left(\nabla \rho_1 \right) \right) - \frac{1}{m} \left(\nabla \theta_0 \right) \left(\nabla \theta_1 \right). \tag{14}$$

If in Equation (14) the "quantum pressure" term $\nabla \left(\rho_0^{-1} \left(\nabla \rho_1 \right) \right) / 4m$ is negligible, as usually assumed, by substitution we obtain

$$\rho_1 = -\frac{\rho_0}{\lambda \rho_0 + \frac{1}{4m} \nabla \left(\rho_0^{-1} \left(\nabla \rho_0 \right) \right)} \left(\left(\partial_t \theta_1 \right) + \frac{1}{m} \left(\nabla \theta_0 \right) \left(\nabla \theta_1 \right) \right), \tag{15}$$

$$0 = \partial_t \left(\frac{\rho_0}{\lambda \rho_0 + \frac{1}{4m} \nabla \left(\rho_0^{-1} \left(\nabla \rho_0 \right) \right)} \left(\left(\partial_t \theta_1 \right) + \frac{1}{m} \left(\nabla \theta_0 \right) \left(\nabla \theta_1 \right) \right) \right)$$

$$+ \nabla \left(\frac{\rho_0}{\lambda \rho_0 + \frac{1}{4m} \nabla \left(\rho_0^{-1} \left(\nabla \rho_0 \right) \right)} \frac{1}{m} \left(\nabla \theta_0 \right) \left(\left(\partial_t \theta_1 \right) + \frac{1}{m} \left(\nabla \theta_0 \right) \left(\nabla \theta_1 \right) \right) - \frac{\rho_0}{m} \left(\nabla \theta_1 \right) \right) \tag{16}$$

with Equation (16) being a Klein–Gordon equation for the field θ_1. Equation (16) can be written in the form

$$\frac{1}{\sqrt{-g}} \partial_\mu \left(\sqrt{-g} g^{\mu\nu} \partial_\nu \theta_1 \right) = 0, \tag{17}$$

where we have introduced the prefactor $1/\sqrt{-g}$ with g being as usual the determinant of the Lorentzian metric $g_{\mu\nu}$. This equation describes an analogue system for a scalar field in curved spacetime as the quantum field θ_1 propagates on a curved geometry with a metric given by

$$\tilde{\lambda} = \lambda + \frac{1}{4m} \rho_0^{-1} \nabla \left(\rho_0^{-1} \left(\nabla \rho_0 \right) \right), \tag{18}$$

$$v_i = \frac{1}{m} \left(\nabla \theta_0 \right)_i, \tag{19}$$

$$\sqrt{-g} = \sqrt{\frac{\rho_0^3}{m^3 \tilde{\lambda}}}, \tag{20}$$

$$g_{tt} = -\sqrt{\frac{\rho_0}{m \tilde{\lambda}}} \left(\frac{\tilde{\lambda} \rho_0}{m} - v^2 \right), \tag{21}$$

$$g_{ij} = \sqrt{\frac{\rho_0}{m \tilde{\lambda}}} \delta_{ij}, \tag{22}$$

$$g_{ti} = -\sqrt{\frac{\rho_0}{m \tilde{\lambda}}} v_i, \tag{23}$$

where the latin indices i and j characterize spatial components.

If the condensate is homogeneous the superfluid velocity vanishes, the coupling is homogeneous in space ($\tilde{\lambda} = \lambda$), the number density ρ_0 is constant in time, and the only relevant behavior of the condensate wave function is in the time-dependent phase θ_0. Furthermore, in this case there is also no need to neglect the quantum pressure term in Equation (14), as it will be handled easily after Fourier-transforming and it will simply introduce a modified dispersion relation—directly derived from the Bogoliubov spectrum.

If the condensate has an initial uniform number density but is not homogeneous—meaning that the initial phase depends on the position—the evolution will introduce inhomogeneities in the density ρ_0, as described by the continuity equation Equation (11), and the initial configuration will be deformed in time. However, as long as $\nabla^2 \theta_0$ is small, also the variations of ρ_0 are small as well: while there is not a nontrivial stationary analogue metric, the scale of the inhomogeneities will define a timescale for which one could safely assume stationarity. Furthermore, the presence of an external potential $V_{ext}(x)$ in the Hamiltonian, via a term of the form $\int dx V_{ext}(x) \phi^\dagger(x) \phi(x)$, would play a role in the dynamical equation for θ_0, leaving invariant those for ρ_0, ρ_1, and θ_1.

3. Time-Dependent Natural Orbitals

The mean-field approximation presented in the previous section is a solid and consistent formulation for studying weakly interacting BEC [22]. It requires, however, that the quantum state has peculiar features which need to be taken into account. In analogue gravity, these assumptions are tacitly considered, but as they play a crucial role for our treatment, we present a discussion of them in some detail to lay down the ground and the formalism in view of next sections.

As is well known [22], the mean-field approximation, consisting in substituting the operator $\phi(x)$ with its expectation value $\langle \phi(x) \rangle$, is strictly valid when the state considered is coherent, meaning it is an eigenstate of the atomic field operator ϕ:

$$\phi(x) |coh\rangle = \langle \phi(x) \rangle |coh\rangle . \tag{24}$$

For states satisfying this equation, the Gross–Pitaevskii Equation (6) is exact (whereas Equation (7) is still a linearized approximation). Note that the coherent states $|coh\rangle$ are not eigenstates of the number operator, but they are rather quantum superpositions of states with different number of atoms. This is necessary because ϕ is an operator that—in the nonrelativistic limit—destroys a particle. We also observe that the notion of coherent state is valid instantaneously, but it may be in general not preserved along the evolution in presence of an interaction.

The redefinition of the field operator, as in Equation (4), provides a description where the physical degrees of freedom are concealed: the new degrees of freedom are not the excited atoms, but the quantized fluctuations over a coherent state. Formally, this is a simple and totally legit redefinition, but for our discussion, we stress that the quanta created by the operator $\delta \phi$ do not have a direct interpretation as atoms.

Given the above discussion, it is useful to remember that coherent states are not the only states to express the condensation, i.e., the fact that a macroscopic number of particles occupies the same state. As it is stated in the Penrose–Onsager criterion for off-diagonal long-range-order [23,24], the condensation phenomenon is best defined considering the properties of the 2-point correlation functions.

The 2-point correlation function is the expectation value on the quantum state of an operator composed of the creation of a particle in a position x after the destruction of a particle in a different position y: $\langle \phi^\dagger(x) \phi(y) \rangle$. As, by definition, the 2-point correlation function is Hermitian, $\overline{\langle \phi^\dagger(y) \phi(x) \rangle} = \langle \phi^\dagger(x) \phi(y) \rangle$, it can always be diagonalized as

$$\left\langle \phi^\dagger(x) \phi(y) \right\rangle = \sum_I \langle N_I \rangle \overline{f_I(x)} f_I(y) , \tag{25}$$

with
$$\int dx \overline{f_I}(x) f_J(x) = \delta_{IJ}. \tag{26}$$

The orthonormal functions f_I, eigenfunctions of the 2-point correlation function, are known as the natural orbitals, and define a complete basis for the 1-particle Hilbert space. In the case of a time-dependent Hamiltonian (or during the dynamics), they are in turn time-dependent. As for the field operator, to simplify the notation, we will not always explicitly write the time dependence of f_I.

The eigenvalues $\langle N_I \rangle$ are the occupation numbers of these wave functions. The sum of these eigenvalues gives the total number of particles in the state (or the mean value, in the case of superposition of quantum states with different number of particles):

$$\langle N \rangle = \sum_I \langle N_I \rangle. \tag{27}$$

The time-dependent orbitals define creation and destruction operators, and consequently the relative number operators (having as expectation values the eigenvalues of the 2-point correlation function):

$$a_I = \int dx \overline{f_I}(x) \phi(x), \tag{28}$$

$$\left[a_I, a_J^\dagger \right] = \delta_{IJ}, \tag{29}$$

$$\left[a_I, a_J \right] = 0, \tag{30}$$

$$N_I = a_I^\dagger a_I. \tag{31}$$

The state is called "condensate" [23] when one of these occupation numbers is macroscopic (comparable with the total number of particles) and the others are small when compared to it.

In the weakly interacting limit, the condensed fraction $\langle N_0 \rangle / \langle N \rangle$ is approximately equal to 1, and the depletion factor $\sum_{I \neq 0} \langle N_I \rangle / \langle N \rangle$ is negligible. This requirement is satisfied by coherent states that define perfect condensates, as the 2-point correlation functions are a product of the mean-field and its conjugate:

$$\langle coh | \phi^\dagger(x) \phi(y) | coh \rangle = \overline{\langle \phi(x) \rangle} \langle \phi(y) \rangle, \tag{32}$$

with
$$f_0(x) = \langle N_0 \rangle^{-1/2} \langle \phi(x) \rangle, \tag{33}$$

$$\langle N_0 \rangle = \int dy \overline{\langle \phi(y) \rangle} \langle \phi(y) \rangle, \tag{34}$$

$$\langle N_{I \neq 0} \rangle = 0. \tag{35}$$

Therefore, in this case, the set of time-dependent orbitals is given by the proper normalization of the mean-field function with a completion that is the basis for the subspace of the Hilbert space orthogonal to the mean-field. The latter set can be redefined arbitrarily, as the only nonvanishing eigenvalue of the 2-point correlation function is the one relative to mean-field function. The fact that there is a nonvanishing macroscopic eigenvalue implies that there is total condensation, i.e., $\langle N_0 \rangle / \langle N \rangle = 1$.

3.1. Time-Dependent Orbitals Formalism

It is important to understand how we can study the condensate state even if we are not considering coherent states and how the description is related to the mean-field approximation. In this framework, we shall see that the mean-field approximation is not a strictly necessary theoretical requirement for analogue gravity.

With respect to the basis of time-dependent orbitals and their creation and destruction operators, we can introduce a new expression for the atomic field operator, projecting it on the sectors of the Hilbert space as

$$\phi(x) = \phi_0(x) + \phi_1(x)$$
$$= f_0(x) a_0 + \sum_I f_I(x) a_I$$
$$= f_0(x) \left(\int dy \overline{f_0}(y) \phi(y) \right) + \sum_{I \neq 0} f_I(x) \left(\int dy \overline{f_I}(y) \phi(y) \right). \quad (36)$$

The two parts of the atomic field operator so defined are related to the previous mean-field $\langle \phi \rangle$ and quantum fluctuation $\delta\phi$ expressions given in Section 2. The standard canonical commutation relation of the background field is of order V^{-1}, where V denotes the volume of the system

$$\left[\phi_0(x), \phi_0^\dagger(y) \right] = f_0(x) \overline{f_0}(y) = \mathcal{O}\left(V^{-1}\right). \quad (37)$$

Note that although the commutator (37) does not vanish identically, it is negligible in the limit of large V.

In the formalism (36), the condensed part of the field is described by the operator ϕ_0 and the orbital producing it through projection, the 1-particle wave function f_0. The dynamics of the function f_0, the 1-particle wave function that best describes the collective behavior of the condensate, can be extracted by using the relations

$$\left\langle \phi^\dagger(x) \phi(y) \right\rangle = \sum_I \langle N_I \rangle \overline{f_I}(x) f_I(y), \quad (38)$$

$$\left\langle \left[a_K^\dagger a_J, H \right] \right\rangle = i \partial_t \langle N_J \rangle \delta_{JK} + i \left(\langle N_K \rangle - \langle N_J \rangle \right) \left(\int dx \overline{f_J}(x) f_K(x) \right) \quad (39)$$

and the evolution of the condensate 1-particle wave function

$$i \partial_t f_0(x) = \left(\int dy \overline{f_0}(y) (i \partial_t f_0(y)) \right) f_0(x) + \sum_{I \neq 0} \left(\int dy \overline{f_0}(y) (i \partial_t f_I(y)) \right) f_I(x)$$
$$= \left(\int dy \overline{f_0}(y) (i \partial_t f_0(y)) \right) f_0(x) + \sum_{I \neq 0} \frac{1}{\langle N_0 \rangle - \langle N_I \rangle} \left\langle \left[a_0^\dagger a_I, H \right] \right\rangle f_I(x)$$
$$= -\frac{i}{2} \frac{\partial_t \langle N_0 \rangle}{\langle N_0 \rangle} f_0(x) + \left(-\frac{\nabla^2}{2m} f_0(x) \right) + \frac{1}{\langle N_0 \rangle} \left\langle a_0^\dagger \left[\phi(x), V \right] \right\rangle +$$
$$+ \sum_{I \neq 0} \frac{\langle N_I \rangle \left\langle a_0^\dagger [a_I, V] \right\rangle + \langle N_0 \rangle \left\langle [a_0^\dagger, V] a_I \right\rangle}{\langle N_0 \rangle \left(\langle N_0 \rangle - \langle N_I \rangle \right)} f_I(x) \quad (40)$$

we assumed at any time $\langle N_0 \rangle \neq \langle N_{I \neq 0} \rangle$. The above equation is valid for a condensate when the dynamics is driven by a Hamiltonian operator composed of a kinetic term and a generic potential V, but we are interested in the case of Equation (1). Furthermore, $f_0(x)$ can be redefined through an overall phase transformation, $f_0(x) \to e^{i\Theta} f_0(x)$, with any arbitrary time-dependent real function Θ. We have chosen the overall phase to satisfy the final expression Equation (40), as it is the easiest to compare with the Gross–Pitaevskii Equation (6).

3.2. Connection with the Gross–Pitaevskii Equation

In this section, we discuss the relation between the function f_0—the eigenfunction of the 2-point correlation function with macroscopic eigenvalue—and the solution of the Gross–Pitaevskii equation, approximating the mean-field function for quasi-coherent states. In particular, we aim at comparing the equations describing their dynamics, detailing under which approximations they show the same behavior. This discussion provides a preliminary technical basis for the study of the effect of the quantum correlations between the background condensate and the quasi-particles, which are present when the quantum nature of the condensate field operator is retained and it is not just approximated by a number, as done when performing the standard Bogoliubov approximation. We refer to the work by the authors of [21] for a review on the Bogoliubov approximation in weakly imperfect Bose gases and to the work by the authors of [25] for a presentation of rigorous results on the condensation properties of dilute bosons.

The Gross–Pitaveskii Equation (6) is an approximated equation for the mean-field dynamics. It holds when the backreaction of the fluctuations $\delta\phi$ on the condensate—described by a coherent state—is negligible. This equation includes a notion of number conservation, meaning that the approximation of the interaction term implies that the spatial integral of the squared norm of the solution of the equation is conserved:

$$\left\langle \phi^\dagger(x)\phi(x)\phi(x)\right\rangle \approx \overline{\langle\phi(x)\rangle}\langle\phi(x)\rangle\langle\phi(x)\rangle \tag{41}$$

$$\Downarrow$$

$$i\partial_t \int dx \overline{\langle\phi(x)\rangle}\langle\phi(x)\rangle = 0. \tag{42}$$

This depends on the fact that only the leading term of the interaction is included in the equation. Therefore we can compare the Gross–Pitaevskii equation for the mean-field with the equation for $\langle N_0 \rangle^{1/2} f_0(x)$ approximated to leading order, i.e., $\langle\phi(x)\rangle$ should be compared to the function $f_0(x)$ under the approximation that there is no depletion from the condensate. If we consider the approximations

$$\left\langle a_0^\dagger[\phi(x),V]\right\rangle \approx \lambda \langle N_0\rangle^2 \overline{f_0}(x)f_0(x)f_0(x), \tag{43}$$

$$i\partial_t \langle N_0\rangle = \left\langle [a_0^\dagger a_0, V]\right\rangle \approx 0, \tag{44}$$

$$\sum_{I\neq 0} \frac{\langle N_I\rangle \langle a_0^\dagger[a_I,V]\rangle + \langle N_0\rangle\langle[a_0^\dagger,V]a_I\rangle}{\langle N_0\rangle - \langle N_I\rangle} f_I(x) \approx 0, \tag{45}$$

we obtain that $\langle N_0\rangle^{1/2} f_0(x)$ satisfies the Gross–Pitaevskii equation.

The approximation in Equation (43) is easily justified, as we are retaining only the leading order of the expectation value $\langle a_0^\dagger[\phi,V]\rangle$ and neglecting the others, which depend on the operators ϕ_1 and $\phi_1^\dagger$ and are of order smaller than $\langle N_0\rangle^2$. The second equation Equation (44) is derived from the previous one as a direct consequence, as the depletion of N_0 comes from the subleading terms $\langle a_0^\dagger \phi_1^\dagger \phi_1 a_0\rangle$ and $\langle a_0^\dagger a_0^\dagger \phi_1 \phi_1\rangle$. The first of these two terms is of order $\langle N_0\rangle$, having its main contributions from separable expectation values—$\langle a_0^\dagger \phi_1^\dagger \phi_1 a_0\rangle \approx \langle N_0\rangle\langle\phi_1^\dagger\phi_1\rangle$—and the second is of the same order due to the dynamics. The other terms are even more suppressed, as can be argued considering that they contain an odd number of operators ϕ_1. Taking their time derivatives, we observe that they arise from the second order in the interaction, making these terms negligible in the regime of weak interaction.

The terms $\langle a_0^\dagger a_0^\dagger a_0 \phi_1\rangle$ are also subleading with respect to those producing the depletion, since the separable contributions—$\langle a_0^\dagger a_0\rangle \langle a_0^\dagger \phi_1\rangle$—vanish by definition, and the remaining describe the correlation between small operators, acquiring relevance only while the many-body quantum state is mixed by the depletion of the condensate:

$$\left\langle a_0^\dagger a_0^\dagger a_0 \phi_1\right\rangle = \left\langle a_0^\dagger \phi_1 (N_0 - \langle N_0\rangle)\right\rangle. \tag{46}$$

Using the same arguments we can assume the approximation in Equation (45) to hold, as the denominator of order $\langle N_0 \rangle$ is sufficient to suppress the terms in the numerator, which are negligible with respect to the leading term in Equation (43).

The leading terms in Equation (45) do not affect the depletion of N_0, but they may be of the same order. They depend on the expectation value

$$\left\langle a_0^\dagger [a_I, V] \right\rangle \approx \lambda \left(\int dx \overline{f_I}(x) \overline{f_0}(x) f_0(x) f_0(x) \right) \langle N_0 \rangle^2 . \tag{47}$$

Therefore, these terms with the mixed action of ladder operators relative to the excited part and the condensate are completely negligible when the integral $\int \overline{f_I} \overline{f_0} f_0 f_0$ is sufficiently small. This happens when the condensed 1-particle state is approximately $f_0 \approx V^{-1/2} e^{i\theta_0}$, i.e., when the atom number density of the condensate is approximately homogeneous.

Moreover, in many cases of interest, it often holds that the terms in the LHS of Equation (45) vanish identically: if the quantum state is an eigenstate of a conserved charge, e.g., total momentum or total angular momentum, the orbitals must be labeled with a specific value of charge. The relative ladder operators act by adding or removing from the state such charge, and for any expectation value not to vanish the charges must cancel out. In the case of homogeneity of the condensate and translational invariance of the Hamiltonian, this statement regards the conservation of momentum. In particular, if the state is invariant under translations, we have

$$\left\langle a_{k_1}^\dagger a_{k_2}^\dagger a_{k_3} a_{k_4} \right\rangle = \delta_{k_1+k_2,k_3+k_4} \left\langle a_{k_1}^\dagger a_{k_2}^\dagger a_{k_3} a_{k_4} \right\rangle , \tag{48}$$

$$\left\langle a_0^\dagger a_0^\dagger a_0 a_k \right\rangle = 0 \tag{49}$$

(for $k \neq 0$).

In conclusion, we obtain that for a condensate with a quasi-homogeneous density a good approximation for the dynamics of the function $\langle N_0 \rangle^{1/2} f_0$, the rescaled 1-particle wave function macroscopically occupied by the condensate, is provided by

$$\begin{aligned} i\partial_t \left(\langle N_0 \rangle^{1/2} f_0(x) \right) &= -\tfrac{\nabla^2}{2m} \left(\langle N_0 \rangle^{1/2} f_0(x) \right) + \lambda \langle N_0 \rangle^{3/2} \overline{f_0}(x) f_0(x) f_0(x) \\ &+ \lambda \langle N_0 \rangle^{-1} \left(\langle a_0^\dagger \phi_1(x) \phi_1^\dagger(x) a_0 \rangle + \langle a_0^\dagger \phi_1^\dagger(x) \phi_1(x) a_0 \rangle \right) \left(\langle N_0 \rangle^{1/2} f_0(x) \right) \\ &+ \lambda \langle N_0 \rangle^{-1} \langle a_0^\dagger \phi_1(x) a_0^\dagger \phi_1(x) \rangle \left(\langle N_0 \rangle^{1/2} \overline{f_0}(x) \right) + \mathcal{O} \left(\lambda \langle N_0 \rangle^{1/2} V^{-3/2} \right) . \end{aligned} \tag{50}$$

This equation is equivalent to the Gross–Pitaveskii equation Equation (6) when we consider only the leading terms, i.e., the first line of Equation (50). If we also consider the remaining lines of the Equation (50), i.e., if we include the effect of the depletion, we obtain an equation that should be compared to the equation for the mean-field function up to the terms quadratic in the operators $\delta\phi$. The two equations are analogous when making the identification:

$$\langle N_0 \rangle^{1/2} f_0 \sim \langle \phi \rangle , \tag{51}$$

$$\left\langle \phi_0^\dagger \phi_1^\dagger \phi_1 \phi_0 \right\rangle \sim \overline{\langle \phi \rangle} \langle \phi \rangle \left\langle \delta\phi^\dagger \delta\phi \right\rangle , \tag{52}$$

$$\left\langle \phi_0^\dagger \phi_0^\dagger \phi_1 \phi_1 \right\rangle \sim \overline{\langle \phi \rangle \langle \phi \rangle} \left\langle \delta\phi \delta\phi \right\rangle . \tag{53}$$

The possible ambiguities in comparing the two equations come from the arbitrariness in fixing the overall time-dependent phases of the functions f_0 and $\langle \phi \rangle$, and from the fact that the commutation relations for the operators ϕ_1 and the operators $\delta\phi$ differ from each other by a term going as $\overline{f_0} f_0$, as seen in Equation (37). This causes an apparent difference when comparing the two terms:

$$\lambda \langle N_0 \rangle^{-1} \left(\langle a_0^\dagger \phi_1(x) \phi_1^\dagger(x) a_0 \rangle + \langle a_0^\dagger \phi_1^\dagger(x) \phi_1(x) a_0 \rangle \right) \left(\langle N_0 \rangle^{1/2} f_0(x) \right) \sim 2\lambda \left\langle \delta\phi^\dagger(x) \delta\phi(x) \right\rangle \langle \phi(x) \rangle . \tag{54}$$

However, the difference can be reabsorbed—manipulating the RHS—in a term which only affects the overall phase of the mean-field, not the superfluid velocity.

The equation for the depletion can be easily derived for the number-conserving approach and compared to the result in the Bogoliubov approach. As seen, the dynamical equation for f_0 contains the information for the time derivative of its occupation number. Projecting the derivative along the function itself and taking the imaginary part, one gets

$$\frac{1}{2}i\partial_t \langle N_0 \rangle = \frac{1}{2}\left\langle \left[a_0^\dagger a_0, V\right]\right\rangle$$
$$= i\mathrm{Im}\left(\int \mathrm{d}x \, \langle N_0 \rangle^{1/2} \overline{f_0}(x) \, i\partial_t \left(\langle N_0 \rangle^{1/2} f_0(x)\right)\right)$$
$$\approx \frac{\lambda}{2}\left(\int \mathrm{d}x \left\langle \phi_0^\dagger(x)\phi_0^\dagger(x)\phi_1(x)\phi_1(x)\right\rangle - \left\langle \phi_1^\dagger(x)\phi_1^\dagger(x)\phi_0(x)\phi_0(x)\right\rangle\right). \quad (55)$$

We can now compare this equation to the one for the depletion in mean-field description

$$\frac{1}{2}i\partial_t N = \frac{1}{2}i\partial_t \left(\int \mathrm{d}x \overline{\langle \phi(x) \rangle} \langle \phi(x) \rangle\right)$$
$$= i\mathrm{Im}\left(\int \overline{\langle \phi(x) \rangle} i\partial_t \langle \phi(x) \rangle\right)$$
$$\approx \frac{\lambda}{2}\left(\int \overline{\langle \phi(x) \rangle \langle \phi(x) \rangle} \langle \delta\phi(x)\delta\phi(x)\rangle - \langle \phi(x) \rangle \langle \phi(x) \rangle \left\langle \delta\phi^\dagger(x)\delta\phi^\dagger(x)\right\rangle\right). \quad (56)$$

The two expressions are consistent with each other

$$\left\langle \phi_0^\dagger(x)\phi_0^\dagger(x)\phi_1(x)\phi_1(x)\right\rangle \sim \overline{\langle \phi(x) \rangle \langle \phi(x) \rangle} \langle \delta\phi(x)\delta\phi(x)\rangle. \quad (57)$$

For coherent states, one expects to find equivalence between $\delta\phi$ and ϕ_1. To do so, we need to review the number-conserving formalism that can provide the same description used for analogue gravity in the general case, e.g., when there is a condensed state with features different from those of coherent states.

4. Number-Conserving Formalism

Within the mean-field framework, the splitting of the field obtained by translating the field operator ϕ by the mean-field function produces the new field $\delta\phi$. This redefinition of the field also induces a corresponding one of the Fock space of the many-body states which, to a certain extent, hides the physical atom degrees of freedom, as the field $\delta\phi$ describes the quantum fluctuations over the mean-field wave function instead of atoms.

Analogue gravity is defined, considering this field and its Hermitian conjugate, and properly combined, to study the fluctuation of phase. The fact that $\delta\phi$ is obtained by translation provides this field with canonical commutation relation. The mean-field description for condensates holds for coherent states and is a good approximation for quasi-coherent states.

When we consider states with fixed number of atoms, and therefore not coherent states, it is better to consider different operators to study the fluctuations. One can do it following the intuition that the fluctuation represents a shift of a single atom from the condensate to the excited fraction and vice versa. Our main reason here to proceed this way is that we are forced to retain the quantum nature of the condensate. We therefore want to adopt the established formalism of number-conserving ladder operators (see, e.g., the work by the authors of [20]) to obtain a different expression for the Bogoliubov–de Gennes equation, studying the excitations of the condensate in these terms. We can adapt this discussion to the time-dependent orbitals.

An important remark is that the qualitative point of introducing the number-conserving approach is conceptually separated from the fact that higher-order terms are neglected by Bogoliubov approximations anyway [20,26]. Indeed, neglecting the commutation relations for a_0 would always imply the impossibility to describe the correlations between quasi-particles and condensate, even when going beyond the Bogoliubov approximation (e.g., by adding terms with three quasi-particle operators). Including such terms, in a growing level of accuracy (and complexity), the main difference would be that the true quasi-particles of the systems do no longer coincide with the Bogoliubov ones. From a practical point of view, this makes clearly a (possibly) huge quantitative difference for the energy spectrum, correlations between quasi-particles, transport properties and observables. Nevertheless, this does not touch at heart that the quantum nature of the condensate is not retained. A discussion of the terms one can include beyond Bogoliubov approximation, and the resulting hierarchy of approximations, is presented in the work by the authors of [26]. Here, our point is rather of principle, i.e., the investigation of the consequences of retaining the operator nature of a_0. Therefore, we used the standard Bogoliubov approximation, improved via the introduction of number-conserving operators.

If we consider the ladder operators a_I, satisfying by definition the relations in Equations (29)–(30), and keep as reference the state a_0 for the condensate, it is a straightforward procedure to define the number-conserving operators $\alpha_{I \neq 0}$, one for each excited wave function, according to the relations

$$\alpha_I = N_0^{-1/2} a_0^\dagger a_I, \tag{58}$$

$$\left[\alpha_I, \alpha_J^\dagger\right] = \delta_{IJ} \qquad \forall I, J \neq 0, \tag{59}$$

$$\left[\alpha_I, \alpha_J\right] = 0 \qquad \forall I, J \neq 0. \tag{60}$$

The degree relative to the condensate is absorbed into the definition, from the hypothesis of number conservation. These relations hold for $I, J \neq 0$, and obviously there is no number-conserving ladder operator relative to the condensed state. The operators α_I are not a complete set of operators to describe the whole Fock space, but they span any subspace of given number of total atoms. To move from one another it would be necessary to include the operator a_0.

This restriction to a subspace of the Fock space is analogous to what is implicitly done in the mean-field approximation, where one considers the subspace of states which are coherent with respect to the action of the destruction operator associated to the mean-field function.

In this set-up, we need to relate the excited part described by ϕ_1 to the usual translated field $\delta\phi$, and obtain an equation for its dynamics related to the Bogoliubov–de Gennes equation. To do so, we need to study the linearization of the dynamics of the operator ϕ_1, combined with the proper operator providing the number conservation

$$N_0^{-1/2} a_0^\dagger \phi_1 = N_0^{-1/2} a_0^\dagger (\phi - \phi_0)$$
$$= \sum_{I \neq 0} f_I \alpha_I. \tag{61}$$

As long as the approximations needed to write a closed dynamical equation for ϕ_1 are compatible with those under which the equation for the dynamics of f_0 resembles the Gross–Pitaevskii equation, i.e., as long as the time derivative of the operators α_I can be written as a combination of the α_I themselves, we can expect to have a set-up for analogue gravity. In fact, in this case, the functional form of the dynamical equations of the system will allow following the standard steps of the derivation reviewed in Section 2.

Therefore, we consider the order of magnitude of the various contributions to the time derivative of $\left(N_0^{-1/2} a_0^\dagger \phi_1\right)$. We have already discussed the time evolution of the function f_0: from the latter it depends the evolution of the operator a_0, since it is the projection along f_0 of the full field operator ϕ.

At first we observed that the variation in time of N_0 must be of smaller order, both for the definition of condensation and because of the approximations considered in the previous section:

$$i\partial_t N_0 = i\partial_t \left(\int dxdy f_0(x) \overline{f_0}(y) \phi^\dagger(x) \phi(y) \right)$$
$$= \left[a_0^\dagger a_0, V \right] + \sum_{I \neq 0} \frac{1}{\langle N_0 \rangle - \langle N_I \rangle} \left(\left\langle \left[a_0^\dagger a_I, V \right] \right\rangle a_I^\dagger a_0 + \left\langle \left[a_I^\dagger a_0, V \right] \right\rangle a_0^\dagger a_I \right). \qquad (62)$$

Of these terms, the summations are negligible due to the prefactors $\left\langle \left[a_I^\dagger a_0, V \right] \right\rangle$ and their conjugates. Moreover, as $\langle a_I^\dagger a_0 \rangle$ is vanishing, the dominant term is the first, which has contributions of at most the order of the depletion factor.

The same can be argued for both the operators a_0 and $N_0^{-1/2}$: their derivatives do not provide leading terms when we consider the derivative of the composite operator $\left(N_0^{-1/2} a_0^\dagger \phi_1 \right)$, only the derivative of the last operator ϕ_1 being relevant. At leading order, we have

$$i\partial_t \left(N_0^{-1/2} a_0^\dagger \phi_1 \right) \approx i \left(N_0^{-1/2} a_0^\dagger \right) (\partial_t \phi_1). \qquad (63)$$

We, therefore, have to analyze the properties of $\partial_t \phi$, considering the expectation values between the orthogonal components ϕ_0 and ϕ_1 and their time derivatives:

$$\left\langle \phi_0^\dagger(y) (i\partial_t \phi_0(x)) \right\rangle = \left(\langle N_0 \rangle^{1/2} \overline{f_0}(y) \right) i\partial_t \left(\langle N_0 \rangle^{1/2} f_0(x) \right), \qquad (64)$$

$$\left\langle \phi_0^\dagger(y) (i\partial_t \phi_1(x)) \right\rangle = -\sum_{I \neq 0} \overline{f_0}(y) f_I(x) \frac{\langle N_I \rangle \langle a_0^\dagger [a_I, V] \rangle + \langle N_0 \rangle \langle [a_0^\dagger, V] a_I \rangle}{\langle N_0 \rangle - \langle N_I \rangle}$$
$$= -\left\langle \left(i\partial_t \phi_0^\dagger(y) \right) \phi_1(x) \right\rangle, \qquad (65)$$

$$\left\langle \phi_1^\dagger(y) (i\partial_t \phi_1(x)) \right\rangle = \left\langle \phi_1^\dagger(y) \left(-\frac{\nabla^2}{2m} \phi_1(x) \right) \right\rangle + \left\langle \phi_1^\dagger(y) [\phi_1(x), V] \right\rangle$$
$$- \sum_{I \neq 0} \overline{f_I}(y) f_0(x) \frac{\langle N_I \rangle \langle [a_I^\dagger a_0, V] \rangle}{\langle N_0 \rangle - \langle N_I \rangle}. \qquad (66)$$

The first equation shows that the function $\langle N_0 \rangle^{1/2} f_0(x)$ assumes the same role of the solution of Gross–Pitaevskii equation in the mean-field description. As long as the expectation value $\left\langle \left[a_0^\dagger a_I, V \right] \right\rangle$ is negligible, we have that the mixed term described by the second equation is also negligible—as it can be said for the last term in the third equation—so that the excited part ϕ_1 can be considered to evolve separately from ϕ_0 in first approximation. Leading contributions from $\left\langle \phi_1^\dagger [\phi_1, V] \right\rangle$ must be those quadratic in the operators ϕ_1 and $\phi_1^\dagger$, and therefore the third equation can be approximated as

$$\left\langle \phi_1^\dagger i\partial_t \phi_1 \right\rangle \approx \left\langle \phi_1^\dagger \left(-\frac{\nabla^2}{2m} \phi_1 + 2\lambda \phi_0^\dagger \phi_0 \phi_1 + \lambda \phi_1^\dagger \phi_0 \phi_0 \right) \right\rangle. \qquad (67)$$

This equation can be compared to the Bogoliubov–de Gennes equation. If we rewrite it in terms of the number-conserving operators, and we consider the fact that the terms mixing the derivative of ϕ_1 with ϕ_0 are negligible, we can write an effective linearized equation for $N_0^{-1/2} a_0^\dagger \phi_1$:

$$i\partial_t \left(N_0^{-1/2} a_0^\dagger \phi_1(x) \right) \approx -\frac{\nabla^2}{2m} \left(N_0^{-1/2} a_0^\dagger \phi_1(x) \right) + 2\lambda \rho_0(x) \left(N_0^{-1/2} a_0^\dagger \phi_1(x) \right)$$
$$+ \lambda \rho_0(x) e^{2i\theta_0(x)} \left(\phi_1^\dagger(x) a_0 N_0^{-1/2} \right). \qquad (68)$$

In this equation, we use the functions ρ_0 and θ_0, which are obtained from the condensed wave function, by writing it as $\langle N_0 \rangle^{1/2} f_0 = \rho_0^{1/2} e^{i\theta_0}$. One can effectively assume the condensed function to be the solution of the Gross–Pitaevskii equation, as the first corrections will be of a lower power of $\langle N_0 \rangle$ (and include a backreaction from this equation itself).

Assuming that ρ_0 is, at first approximation, homogeneous, implies that the term $\langle [a_0^\dagger a_I, V] \rangle$ is negligible. If ρ_0 and θ_0 are ultimately the same as those obtained from the Gross–Pitaevskii equation, the same equation that holds for the operator $\delta\phi$ can be assumed to hold for the operator $N_0^{-1/2} a_0^\dagger \phi_1$. The solution for the mean-field description of the condensate is therefore a general feature of the system in studying the quantum perturbation of the condensate, not strictly reserved to coherent states.

Although having strongly related dynamical equations, the substantial difference between the operators $\delta\phi$ of Equation (4) and $N_0^{-1/2} a_0^\dagger \phi_1$ is that the number-conserving operator does not satisfy the canonical commutation relations with its Hermitian conjugate, as we have extracted the degree of freedom relative to the condensed state

$$\left[\phi_1(x), \phi_1^\dagger(y)\right] = \delta(x,y) - f_0(x) \overline{f_0}(y) . \tag{69}$$

Although this does not imply a significant obstruction, one must remind that the field ϕ_1 should never be treated as a canonical quantum field. What has to be done, instead, is considering its components with respect to the basis of time-dependent orbitals. Each mode of the projection ϕ_1 behaves as if it is a mode of a canonical scalar quantum field in a curved spacetime. Keeping this in mind, we can safely retrieve analogue gravity.

5. Analogue Gravity with Atom Number Conservation

In the previous section, we discussed the equivalent of the Bogoliubov–de Gennes equation in a number-conserving framework. Our aim in the present Section is to extend this description to analogue gravity.

The field operators required for analogue gravity will differ from those relative to the mean-field description, and they should be defined considering that we have by construction removed the contribution from the condensed 1-particle state f_0. The dynamical equation for the excited part in the number-conserving formalism Equation (68) appears to be the same as for the case of coherent states (as in Equation (7)), but instead of the field $\delta\phi$ one has $N_0^{-1/2} a_0^\dagger \phi_1$, where we remind that $N_0 = a_0^\dagger a_0$.

Using the Madelung representation, we may redefine the real functions ρ_0 and θ_0 from the condensed wave function f_0 and the expectation value $\langle N_0 \rangle$. Approximating at the leading order, we can obtain their dynamics as in the quantum Euler Equations (11)–(12):

$$\langle N_0 \rangle^{1/2} f_0 = \rho_0^{1/2} e^{i\theta_0} . \tag{70}$$

These functions enter in the definition of the quantum operators θ_1 and ρ_1, which take a different expression from the usual Madelung representation when we employ the set of number-conserving ladder operators

$$\begin{aligned}
\theta_1 &= -\frac{i}{2} \langle N_0 \rangle^{-1/2} \sum_{I \neq 0} \left(\frac{f_I}{f_0} N_0^{-1/2} a_0^\dagger a_I - \frac{\overline{f_I}}{\overline{f_0}} a_I^\dagger a_0 N_0^{-1/2} \right) \\
&= -\frac{i}{2} \left(\frac{N_0^{-1/2} \phi_0^\dagger \phi_1 - \phi_1^\dagger \phi_0 N_0^{-1/2}}{\langle N_0 \rangle^{1/2} \overline{f_0} f_0} \right) ,
\end{aligned} \tag{71}$$

$$\begin{aligned}
\rho_1 &= \langle N_0 \rangle^{1/2} \sum_{I \neq 0} \left(\overline{f_0} f_I N_0^{-1/2} a_0^\dagger a_I + f_0 \overline{f_I} a_I^\dagger a_0 N_0^{-1/2} \right) = \\
&= \langle N_0 \rangle^{1/2} \left(N_0^{-1/2} \phi_0^\dagger \phi_1 + \phi_1^\dagger \phi_0 N_0^{-1/2} \right) .
\end{aligned} \tag{72}$$

From Equations (71) and (72), we observe that the structure of the operators θ_1 and ρ_1 consists of a superposition of modes, each dependent on a different eigenfunction f_I of the 2-point correlation function, with a sum over the index $I \neq 0$.

The new fields θ_1 and ρ_1 do not satisfy the canonical commutation relations since the condensed wave function f_0 is treated separately by definition. However, these operators could be analyzed mode-by-mode, and therefore be compared in full extent to the modes of quantum fields in curved spacetime to which they are analogous. Their modes satisfy the relations

$$[\theta_I, \rho_J] = -i\overline{f_I} f_I \delta_{JI} \qquad \forall I, J \neq 0. \tag{73}$$

Equation (73) is a basis-dependent expression, which can, in general, be found for the fields of interest. In the simplest case of homogeneous density of the condensate ρ_0, this commutation relations reduce to $-i\delta_{IJ}$, and the Fourier transform provides the tools to push the description to full extent where the indices labeling the functions are the momenta k.

The equations for analogue gravity are found under the usual assumptions regarding the quantum pressure, i.e., the space gradients of the atom densities are assumed to be small. When considering homogeneous condensates this requirement is of course satisfied. In nonhomogeneous condensates, we require

$$\nabla\left(\rho_0^{-1}(\nabla\rho_0)\right) \ll 4m\lambda\rho_0, \tag{74}$$

$$\nabla\left(\rho_0^{-1}(\nabla\rho_1)\right) \ll 4m\lambda\rho_1. \tag{75}$$

Making the first assumption (74), the effective coupling constant $\tilde{\lambda}$ is a global feature of the system with no space dependence. This means that all the inhomogeneities of the system are encoded in the velocity of the superfluid, the gradient of the phase of the condensate. As stated before, the continuity equation can induce inhomogeneities in the density if there are initial inhomogeneities in the phase, but for sufficiently short intervals of time, the assumption is satisfied. Another effect of the first assumption (74) is that the term $\int dx \overline{f_I} f_0 f_0 f_0$ is negligible. The more ρ_0 is homogeneous, the closer this integral is to vanishing, making the description more consistent. The second assumption (75) is a general requirement in analogue gravity, needed to have local Lorentz symmetry, and therefore a proper Klein–Gordon equation for the field θ_1. When ρ_0 is homogeneous, this approximation means considering only small momenta, for which we have the usual dispersion relation.

Under these assumptions, the usual equations for analogue gravity are obtained:

$$\rho_1 = -\frac{1}{\lambda}\left((\partial_t \theta_1) + \frac{1}{m}(\nabla\theta_0)(\nabla\theta_1)\right), \tag{76}$$

$$(\partial_t \rho_1) = -\frac{1}{m}\nabla\left(\rho_1(\nabla\theta_0) + \rho_0(\nabla\theta_1)\right) \tag{77}$$

$$\Downarrow$$

$$0 = \partial_t\left(-\frac{1}{\lambda}(\partial_t\theta_1) - \frac{\delta^{ij}}{m\lambda}(\nabla_j\theta_0)(\nabla_i\theta_1)\right)$$
$$+ \nabla_j\left(-\frac{\delta^{ij}}{m\lambda}(\nabla_i\theta_0)(\partial_t\theta_1) + \left(\frac{\delta^{ij}\rho_0}{m} - \frac{1}{\lambda}\frac{\delta^{il}}{m}\frac{\delta^{jm}}{m}(\nabla_l\theta_0)(\nabla_m\theta_0)\right)(\nabla_i\theta_1)\right) \tag{78}$$

so that θ_1 is the analogue of a scalar massless field in curved spacetime. However, the operator θ_1 is intrinsically unable to provide an exact full description of a massless field since it is missing the mode f_0. Therefore, the operator θ_1 is best handled when considering the propagation of its constituent modes, and relating them to those of the massless field.

The viability of this description as a good analogue gravity set-up is ensured, ultimately, by the fact that the modes of θ_1, i.e., the operators describing the excited part of the atomic field, have a closed dynamics. The most important feature in the effective dynamics of the number-conserving operators $N_0^{-1/2} a_0^\dagger a_I$, as described in equation Equation (68), is that its time derivative can be written as a composition of the same set of number operators, and this enables the analogue model.

In the following, we continue with the case of an homogenous condensate, which is arguably the most studied case in analogue gravity. The description is enormously simplified by the fact that the gradient of the condensed wave function vanishes, since $f_0 = V^{-1/2}$, meaning that the condensed state is fully described by the state of null momentum $k = 0$. In a homogeneous BEC, all the time-dependent orbitals are labeled by the momenta they carry, and at every moment in time, we can apply the same Fourier transform to transform the differential equations in the space of coordinates to algebraic equations in the space of momenta. We expect that the number-conserving treatment of the inhomogenous condensate follows along the same lines, albeit being technically more complicated.

6. Simulating Cosmology in Number-Conserving Analogue Gravity

With an homogeneous condensate we can simulate a cosmology with a scale factor changing in time—as long as we can control and modify in time the strength of the 2-body interaction λ—and we can verify the prediction of quantum field theory in curved spacetime that in an expanding universe one should observe a cosmological particle creation [9,27–30]. In this set-up, there is no ambiguity in approximating the mixed term of the interaction potential, as discussed in the Section 3.2.

To further proceed, we apply the usual transformation to pass from the Bogoliubov description of the atomic system to the set-up of analogue gravity, and we then proceed considering number-conserving operators. It is convenient to adopt a compact notation for the condensate wave function and its approximated dynamics, as discussed previously in Equations (11)–(12) and in Equation (70).

$$f_0(x) \langle N_0 \rangle^{1/2} \equiv \phi_0 = \rho_0^{1/2} e^{i\theta_0}, \tag{79}$$

$$\partial_t \rho_0 = 0, \tag{80}$$

$$\partial_t \theta_0 = -\lambda \rho_0. \tag{81}$$

To study the excitations described by the operator $\theta_1(x)$ we need the basis of time-dependent orbitals, which in the case of a homogeneous condensate is given by the plane waves, the set of orthonormal functions which define the Fourier transform and are labeled by the momenta. By Fourier transforming the operator $\phi_1(x)$, orthogonal to the condensate wave function, we have

$$\delta\phi_k \equiv \int \frac{dx}{\sqrt{V}} e^{-ikx} N_0^{-1/2} a_0^\dagger \phi_1(x)$$
$$= \int \frac{dx}{\sqrt{V}} e^{-ikx} N_0^{-1/2} a_0^\dagger \sum_{q \neq 0} \frac{e^{iqx}}{\sqrt{V}} a_q$$
$$= N_0^{-1/2} a_0^\dagger a_k, \tag{82}$$

$$\left[\delta\phi_k, \delta\phi_{k'}^\dagger\right] = \delta_{k,k'} \qquad \forall k, k' \neq 0. \tag{83}$$

Notice that with the notation of Equation (58), $\delta\phi_k$ would be just α_k and one sees the dependence on the condensate operator a_0.

Following the same approach discussed in the Section 5, we define θ_k and ρ_k. These number-conserving operators are labeled with a nonvanishing momentum and act in the atomic Fock space, in a superposition of two operations, extracting momentum k from the state or introducing momentum $-k$ to it. All the following relations are defined for $k, k' \neq 0$.

$$\theta_k = -\frac{i}{2} \left(\frac{\delta\phi_k}{\phi_0} - \frac{\delta\phi^\dagger_{-k}}{\overline{\phi_0}} \right), \tag{84}$$

$$\rho_k = \rho_0 \left(\frac{\delta\phi_k}{\phi_0} + \frac{\delta\phi^\dagger_{-k}}{\overline{\phi_0}} \right), \tag{85}$$

$$[\theta_k, \rho_{k'}] = -i \left[\delta\phi_k, \delta\phi^\dagger_{-k'} \right] = -i\delta_{k,-k'}, \tag{86}$$

$$\left\langle \delta\phi^\dagger_k(t) \delta\phi_{k'}(t) \right\rangle = \delta_{k,k'} \langle N_k \rangle. \tag{87}$$

Again, we remark that these definitions of θ_k and ρ_k do not provide, through an inverse Fourier transform of these operators, a couple of conjugate real fields, $\theta_1(x)$ and $\rho_1(x)$, with the usual commutation relations as in Equation (10), because they are not relative to a set of functions that form a complete basis of the 1-particle Hilbert space, as the mode $k = 0$ is not included. However, these operators, describing each mode with $k \neq 0$, can be studied separately and they show the same behavior of the components of a quantum field in curved spacetime: the commutation relations in Equation (86) are the same as those that are satisfied by the components of a quantum scalar field.

From the Bogoliubov–de Gennes Equation (68), we get the two coupled dynamical equations for θ_k and ρ_k:

$$\partial_t \theta_k = -\frac{1}{2} \left(\frac{k^2}{2m} + 2\lambda\rho_0 \right) \frac{\rho_k}{\rho_0}, \tag{88}$$

$$\partial_t \frac{\rho_k}{\rho_0} = \frac{k^2}{m} \theta_k. \tag{89}$$

Combining these gives the analogue Klein–Gordon equation for each mode $k \neq 0$:

$$\partial_t \left(-\frac{1}{\lambda\rho_0 + \frac{k^2}{4m}} (\partial_t \theta_k) \right) = \frac{k^2}{m} \theta_k. \tag{90}$$

In this equation, the term due to quantum pressure is retained for convenience, as the homogeneity of the condensed state makes it easy to maintain it in the description. It modifies the dispersion relation and breaks Lorentz symmetry, but the usual expression is found in the limit $\frac{k^2}{2m} \ll 2\lambda\rho_0$.

When the quantum pressure is neglected, the analogue metric tensor is

$$g_{\mu\nu}dx^\mu dx^\nu = \sqrt{\frac{\rho_0}{m\lambda}} \left(-\frac{\lambda\rho_0}{m} dt^2 + \delta_{ij} dx^i dx^j \right). \tag{91}$$

This metric tensor is clearly analogous to that of a cosmological spacetime, where the evolution is given by the time dependence of the coupling constant λ. This low-momenta limit is the regime in which we are mostly interested, because when these conditions are realized the quasi-particles, the excitations of the field θ_k, behave most similarly to particles in a curved spacetime with local Lorentz symmetry.

6.1. Cosmological Particle Production

We now consider a set-up for which the coupling constant varies from an initial value λ to a final value λ' through a transient phase. λ is assumed asymptotically constant for both $t \to \pm\infty$. This set-up has been studied in the Bogoliubov approximation in the works by the authors of [9,27–30] and can be experimentally realized with. e.g., via Feshbach resonance. For one-dimensional Bose gases where significant corrections to the Bogoliubov approximation are expected far from the weakly interacting limit, a study of the large time evolution of correlations was presented in the work by the authors of [31]. Here, our aim is to study the effect of the variation of the coupling constant in the number-conserving framework.

There will be particle creation and the field in general takes the expression

$$\theta_k(t) = \frac{1}{\mathcal{N}_k(t)} \left(e^{-i\Omega_k(t)} c_k + e^{i\Omega_{-k}(t)} c^\dagger_{-k} \right), \qquad (92)$$

where the operators c_k are the creation and destruction operators for the quasi-particles at $t \to -\infty$. For the time $t \to +\infty$, there will be a new set of operators c',

$$\theta_k(t \to -\infty) = \frac{1}{\mathcal{N}_k} \left(e^{-i\omega_k t} c_k + e^{i\omega_k t} c^\dagger_{-k} \right), \qquad (93)$$

$$\theta_k(t \to +\infty) = \frac{1}{\mathcal{N}'_k} \left(e^{-i\omega'_k t} c'_k + e^{i\omega'_k t} c'^\dagger_{-k} \right). \qquad (94)$$

From these equations, in accordance with Equation (88), we obtain

$$\rho_k = -\frac{2\rho_0}{\frac{k^2}{2m} + 2\lambda\rho_0} \partial_t \theta_k \qquad (95)$$

and the two following asymptotic expressions for ρ_k,

$$\rho_k(t \to -\infty) = \frac{2i\omega_k \rho_0}{\frac{k^2}{2m} + 2\lambda\rho_0} \frac{1}{\mathcal{N}_k} \left(e^{-i\omega_k t} c_k - e^{i\omega_k t} c^\dagger_{-k} \right), \qquad (96)$$

$$\rho_k(t \to +\infty) = \frac{2i\omega'_k \rho_0}{\frac{k^2}{2m} + 2\lambda'\rho_0} \frac{1}{\mathcal{N}'_k} \left(e^{-i\omega'_k t} c'_k - e^{i\omega'_k t} c'^\dagger_{-k} \right). \qquad (97)$$

With the previous expressions for θ_k and ρ_k and imposing the commutation relations in Equation (86), we retrieve the energy spectrum $\omega_k = \sqrt{\frac{k^2}{2m}\left(\frac{k^2}{2m} + 2\lambda\rho_0\right)}$ as expected and the (time-dependent) normalization prefactor $\mathcal{N}$:

$$\mathcal{N}_k = \sqrt{4\rho_0 \sqrt{\frac{\frac{k^2}{2m}}{\frac{k^2}{2m} + 2\lambda\rho_0}}}. \qquad (98)$$

The expected commutation relations for the operators c and c' are found (again not including the mode $k = 0$):

$$0 = \left[c_k, c_{k'}\right] = \left[c'_k, c'_{k'}\right], \qquad (99)$$

$$\delta_{k,k'} = \left[c_k, c^\dagger_{k'}\right] = \left[c'_k, c'^\dagger_{k'}\right]. \qquad (100)$$

It is found

$$c'_k = \cosh\Theta_k c_k + \sinh\Theta_k e^{i\varphi_k} c^\dagger_{-k} \qquad (101)$$

with

$$\cosh\Theta_k = \cosh\Theta_{-k}, \qquad (102)$$
$$\sinh\Theta_k e^{i\varphi_k} = \sinh\Theta_{-k} e^{i\varphi_{-k}}. \qquad (103)$$

The initial state in which we are interested is the vacuum of quasi-particles, so that each quasi-particle destruction operators c_k annihilates the initial state (To make contact with the standard

Bogoliubov approximation, if there one denotes by γ_k the quasi-particles one has that the γ_k are a combination of the atom operators a_k, a_{-k} of the form $\gamma_k = u_k a_k + v_k a^\dagger_{-k}$ [20]. Correspondingly, in the number-conserving formalism the quasi-particle operators c_k are a combination of the atom operators $\delta\phi_k \equiv \alpha_k$, $\delta\phi_{-k} \equiv \alpha_{-k}$):

$$c_k \left|in\right\rangle \equiv 0 \qquad \forall k \neq 0. \tag{104}$$

To realize this initial condition, we should impose constraints, in principle, on every correlation function. We focus on the 2-point correlation functions $\langle \delta\phi^\dagger \delta\phi \rangle$ and $\langle \delta\phi \delta\phi \rangle$. In particular, the first of the two determines the number of atoms with momentum k in the initial state:

$$\left\langle \delta\phi^\dagger_k \delta\phi_k \right\rangle = \left\langle a^\dagger_k a_0 N_0^{-1} a^\dagger_0 a_k \right\rangle = \left\langle a^\dagger_k a_k \right\rangle = \langle N_k \rangle. \tag{105}$$

In order for the state to be condensed with respect to the state with momentum 0, it must be that $\langle N_k \rangle \ll \langle N_0 \rangle = \rho_0 V$. When the vacuum condition Equation (104) holds, the 2-point correlation functions can be easily evaluated to be

$$\left\langle \delta\phi^\dagger_k \delta\phi_{k'} \right\rangle = \left(\frac{1}{2} \frac{\frac{k^2}{2m} + \lambda\rho_0}{\sqrt{\frac{k^2}{2m}\left(\frac{k^2}{2m} + 2\lambda\rho_0\right)}} - \frac{1}{2} \right) \delta_{k,k'}$$

$$\approx \frac{1}{4}\sqrt{\frac{2\lambda\rho_0}{\frac{k^2}{2m}}} \delta_{k,k'}, \tag{106}$$

$$\left\langle \delta\phi_{-k} \delta\phi_{k'} \right\rangle = -\frac{e^{2i\theta_0}}{4} \frac{2\lambda\rho_0}{\sqrt{\frac{k^2}{2m}\left(\frac{k^2}{2m} + 2\lambda\rho_0\right)}} \delta_{k,k'}$$

$$\approx -e^{2i\theta_0} \left\langle \delta\phi^\dagger_k \delta\phi_{k'} \right\rangle, \tag{107}$$

where in the last line we have used $\frac{k^2}{2m} \ll 2\lambda\rho_0$, the limit in which the quasi-particles propagate in accordance with the analogue metric Equation (91), and one has to keep into account that the phase of the condensate is time dependent and consequently the last correlator is oscillating.

We now see that the conditions of condensation $\langle N_k \rangle \ll \langle N_0 \rangle$ and of low-momenta translate into

$$\frac{2\lambda\rho_0}{16\langle N_0 \rangle^2} \ll \frac{k^2}{2m} \ll 2\lambda\rho_0. \tag{108}$$

The range of momenta that should be considered is, therefore, set by the number of condensate atoms, the physical dimension of the atomic system, and the strength of the 2-body interaction.

The operators θ_k satisfying Equation (90)—describing the excitations of quasi-particles over a BEC—are analogous to the components of a scalar quantum field in a cosmological spacetime. In particular, if we consider a cosmological metric given in the usual form of

$$g_{\mu\nu} dx^\mu dx^\nu = -d\tau^2 + a^2 \delta_{ij} dx^i dx^j, \tag{109}$$

the analogy is realized for a specific relation between the coupling $\lambda(t)$ and the scale factor $a(\tau)$, which then induces the relation between the laboratory time t and the cosmological time τ. These relations are given by

$$a(\tau(t)) = \left(\frac{\rho_0}{m\lambda(t)}\right)^{1/4} \frac{1}{C}, \quad (110)$$

$$d\tau = \frac{\rho_0}{ma(\tau(t))} \frac{1}{C^2} dt, \quad (111)$$

for an arbitrary constant C.

In cosmology, the evolution of the scale factor leads to the production of particles by cosmological particle creation, as implied by the Bogoliubov transformation relating the operators which, at early and late times, create and destroy the quanta we recognize as particles. The same happens for the quasi-particles over the condensate, as discussed in Section 6, because the coupling λ is time-dependent and the definition itself of quasi-particles changes from initial to final time. The ladder operators associated to these quasi-particles are related to each other by the Bogoliubov transformation introduced in Equation (101), fully defined by the parameters Θ_k and φ_k (which must also satisfy Equations (102) and (103)).

6.2. Scattering Operator

The exact expressions of Θ_k and φ_k depend on the behavior of $\lambda(t)$, which is a function of the cosmological scale parameter, and is therefore different for each cosmological model. They can in general be evaluated with the well-established methods used in quantum field theory in curved spacetimes [32]. In general, it is found that $\cosh\Theta_k > 1$, as the value $\cosh\Theta_k = 1$ (i.e., $\sinh\Theta_k = 0$) is restricted to the case in which λ is a constant for the whole evolution, and the analogue spacetime is simply flat.

The unitary operator describing the evolution from initial to final time is $U(t_{out}, t_{in})$ when $t_{out} \to +\infty$ and $t_{in} \to -\infty$, and the operator, U, is the scattering operator, S. This is exactly the operator acting on the quasi-particles, defining the Bogoliubov transformation in which we are interested

$$c'_k = S^\dagger c_k S. \quad (112)$$

The behavior of c'_k, describing the quasi-particles at late times, can therefore be understood from the behavior of the initial quasi-particle operators c_k when the expression of the scattering operator is known. In particular, the phenomenon of cosmological particle creation is quantified considering the expectation value of the number operator of quasi-particles at late times in the vacuum state as defined by early times operators [32].

Consider as initial state the vacuum of quasi-particles at early times, satisfying the condition Equation (104). It is analogous to a Minkowski vacuum, and the evolution of the coupling $\lambda(t)$ induces a change in the definition of quasi-particles. We find that, of course, the state is not a vacuum with respect to the final quasi-particles c'. It is

$$S^\dagger c_k^\dagger c_k S = c'^\dagger_k c'_k = \left(\cosh\Theta_k c_k^\dagger + \sinh\Theta_k e^{-i\varphi_k} c_{-k}\right)\left(\cosh\Theta_k c_k + \sinh\Theta_k e^{i\varphi_k} c^\dagger_{-k}\right) \quad (113)$$

and

$$\left\langle S^\dagger c_k^\dagger c_k S \right\rangle = \sinh^2\Theta_k \left\langle c_{-k} c^\dagger_{-k} \right\rangle = \sinh^2\Theta_k > 0. \quad (114)$$

We are interested in the effect that the evolution of the quasi-particles have on the atoms. The system is fully characterized by the initial conditions and the Bogoliubov transformation: we have the initial occupation numbers, the range of momenta which we should consider, and the relation between initial and final quasi-particles.

What is most significant is that the quasi-particle dynamics affects the occupation number of the atoms. Considering that for sufficiently large t we are already in the final regime, the field takes the following values,

$$\delta\phi_k(t \to -\infty) = i\rho_0^{1/2} e^{i\theta_0(t)} \frac{1}{\mathcal{N}_k} \left((\mathcal{F}_k + 1) e^{-i\omega_k t} c_k - (\mathcal{F}_k - 1) e^{i\omega_k t} c^\dagger_{-k} \right) \quad (115)$$

$$\delta\phi_k(t \to +\infty) = i\rho_0^{1/2} e^{i\theta_0(t)} \frac{1}{\mathcal{N}'_k} \left((\mathcal{F}'_k + 1) e^{-i\omega'_k t} c'_k - (\mathcal{F}'_k - 1) e^{i\omega'_k t} c'^\dagger_{-k} \right) \quad (116)$$

where $\mathcal{F}_k \equiv \frac{\omega_k}{\frac{k^2}{2m} + 2\lambda\rho_0}$ and $\mathcal{F}'_k \equiv \frac{\omega'_k}{\frac{k^2}{2m} + 2\lambda'\rho_0}$, with $\omega'_k = \sqrt{\frac{k^2}{2m}\left(\frac{k^2}{2m} + 2\lambda'\rho_0\right)}$. One finds

$$\left\langle \delta\phi^\dagger_k(t) \delta\phi_k(t) \right\rangle = \frac{\frac{k^2}{2m} + \lambda'\rho_0}{2\omega'_k} \cosh(2\Theta_k) - \frac{1}{2} + \frac{\lambda'\rho_0 \sinh(2\Theta_k)}{2\omega'_k} \cos(2\omega'_k t - \varphi_k). \quad (117)$$

In Equation (117), the last term is oscillating symmetrically around 0—meaning that the atoms will leave and rejoin the condensate periodically in time—whereas the first two are stationary.

An increase in the value of the coupling λ therefore has deep consequences. It appears explicitly in the prefactor and more importantly it affects the hyperbolic functions $\cosh \Theta_k > 1$, which implies that the mean value is larger than the initial one, differing from the equilibrium value corresponding to the vacuum of quasi-particles.

This result is significant because it explicitly shows that the quasi-particle dynamics influences the underlying structure of atomic particles. Even assuming that the backreaction of the quasi-particles on the condensate is negligible for the dynamics of the quasi-particles themselves, the mechanism of extraction of atoms from the condensate fraction is effective and increases the depletion (as also found in the standard Bogoliubov approach). This extraction mechanism can be evaluated in terms of operators describing the quasi-particles, that can be defined a posteriori, without notion of the operators describing the atoms.

The fact that analogue gravity can be reproduced in condensates independently from the use of coherent states enhances the validity of the discussion. It is not strictly necessary that we have a coherent state to simulate the effects of curvature with quasi-particles, but, in the more general case of condensation, the condensed wave function provides a support for the propagation of quasi-particles. From an analogue gravity point of view, its intrinsic role is that of seeding the emergence of the analogue scalar field [2].

7. Squeezing and Quantum State Structure

The Bogoliubov transformation in Equation (101) leading to the quasi-particle production describes the action of the scattering operator on the ladder operators, relating the operators at early and late times. The linearity of this transformation is obtained by the linearity of the dynamical equation for the quasi-particles, which is particularly simple in the case of homogeneous condensate.

The scattering operator S is unitary by definition, as it is easily checked by its action on the operators c_k. Its full expression can be found from the Bogoliubov transformation, finding the generators of the transformation when the arguments of the hyperbolic functions, the parameters Θ_k, are infinitesimal:

$$S^\dagger c_k S = c'_k = \cosh\Theta_k c_k + \sinh\Theta_k e^{i\varphi_k} c^\dagger_{-k}. \quad (118)$$

It follows

$$S = \exp\left(\frac{1}{2} \sum_{k \neq 0} \left(-e^{-i\varphi_k} c_k c_{-k} + e^{i\varphi_k} c^\dagger_k c^\dagger_{-k}\right) \Theta_k\right). \quad (119)$$

The scattering operator is particularly simple and takes the peculiar expression that is required for producing squeezed states. This is the general functional expression that is found in cosmological particle creation and in its analogue gravity counterparts, whether they are realized in the usual Bogoliubov framework or in its number-conserving reformulation. As discussed previously, the number-conserving formalism is more general, reproduces the usual case when the state is an eigenstate of the destruction operator a_0, and includes the notion that the excitations of the condensate move condensate atoms to the excited part.

The expression in Equation (119) has been found under the hypothesis that the mean value of the operator N_0 is macroscopically larger than the other occupation numbers. Instead of using the quasi-particle ladder operators, S can be rewritten easily in terms of the atom operators. In particular, we remind that the time-independent operators c_k depend on the condensate operator a_0 and can be defined as compositions of number-conserving atom operators $\delta\phi_k(t)$ and $\delta\phi^\dagger_{-k}(t)$ defined in Equation (82). At any time, there can be a transformation from one set of operators to the another. It is significant that the operators c_k commute with the operators $N_0^{-1/2} a_0^\dagger$ and $a_0 N_0^{-1/2}$, which are therefore conserved in time (as long as the linearized dynamics for $\delta\phi_k$ is a good approximation)

$$\left[\delta\phi_k, N_0^{-1/2}(t)\, a_0^\dagger(t)\right] = 0, \tag{120}$$

$$\Downarrow$$

$$\left[c_k, N_0^{-1/2}(t)\, a_0^\dagger(t)\right] = 0, \tag{121}$$

$$\Downarrow$$

$$\left[S, N_0^{-1/2}(t)\, a_0^\dagger(t)\right] = 0. \tag{122}$$

The operator S cannot have other terms apart for those in Equation (119), even if it is defined for its action on the operators c_k, and therefore on a set of functions, which is not a complete basis of the 1-particle Hilbert space. Nevertheless, the notion of number conservation implies its action on the condensate and on the operator a_0.

One could investigate whether it is possible to consider a more general expression with additional terms depending only on a_0 and $a_0^\dagger$, i.e., assuming the scattering operator to be

$$S = \exp\left(\frac{1}{2}\sum_{k\neq 0}\left(Z_k c_k c_{-k} + c_k^\dagger c_{-k}^\dagger Z_{-k}^\dagger\right) + G_0\right), \tag{123}$$

where we could assume that the coefficients of the quasi-particle operators are themselves depending on only a_0 and $a_0^\dagger$, and so G_0. However, the requirement that S commutes with the total number of atoms N implies that so do its generators, and therefore Z and G_0 must be functionally dependent on N_0, and not on a_0 and $a_0^\dagger$ separately, as they do not conserve the total number. Therefore, it must hold that

$$0 = \left[\left(\frac{1}{2}\sum_{k\neq 0}\left(Z_k c_k c_{-k} + c_k^\dagger c_{-k}^\dagger Z_{-k}^\dagger\right) + G_0\right), N\right]. \tag{124}$$

The only expressions in agreement with the linearized dynamical equation for $\delta\phi$ imply that Z and G_0 are multiple of the identity; otherwise, they would modify the evolution of the operators $\delta\phi_k = N_0^{-1/2} a_0^\dagger a_I$, as they do not commute with N_0. This means that that corrections to the scattering operator are possible only involving higher-order corrections (in terms of $\delta\phi$).

The fact that the operator S as in Equation (119) is the only number-conserving operator satisfying the dynamics is remarkable because it emphasizes that the production of quasi-particles is a phenomenon that holds only in terms of excitations of atoms from the condensate to the excited part, with the number of transferred atoms evaluated in the previous subsection. The expression of the scattering operator shows that the analogue gravity system produces states in which the final state presents squeezed quasi-particle states; however, the occurrence of this feature in the emergent dynamics happens only introducing correlations in the condensate, with each quanta of the analogue field θ_1 entangling atoms in the condensate with atoms in the excited part.

The quasi-particle scattering operator obtained in the number-conserving framework is functionally equivalent to that in the usual Bogoliubov description, and the difference between the two appears when considering the atom operators, depending on whether a_0 is treated for its quantum nature or it is replaced with the number $\langle N_0^{1/2} \rangle$. This reflects that the dynamical equations are functionally the same when the expectation value $\langle N_0 \rangle$ is macroscopically large.

There are no requirements on the initial density matrix of the state, and it is not relevant whether the state is a coherent superposition of infinite states with different number of atoms or it is a pure state with a fixed number of atoms in the same 1-particle state. The quasi-particle description holds the same and it provides the same predictions. This is useful for implementing analogue gravity systems, but also a strong hint in interpreting the problem of information loss. When producing quasi-particles in analogue gravity one can, in first approximation, reconstruct the initial expectation values of the excited states and push the description to include the backreaction on the condensate. What we are intrinsically unable to do is reconstruct the entirety of the initial atom quantum state, i.e., how the condensate is composed.

We know that in analogue gravity the evolution is unitary, the final state is uniquely determined by the initial state. Knowing all the properties of the final state we could reconstruct the initial state, and yet the intrinsic inability to infer all the properties of the condensate atoms from the excited part shows that the one needs to access the full correlation properties of the condensate atoms with the quasi-particles to fully appreciate (and retrieve) the unitarity of the evolution.

7.1. Correlations

In the previous section, we made the standard choice of considering as initial state the quasi-particle vacuum. To characterize it with respect to the atomic degrees of freedom, the quasi-particle ladders operators have to be expressed as compositions of the number-conserving atomic operators, manipulating Equations (84) and (92).

By definition, at any time, both sets of operators satisfy the canonical commutation relations (87) and (100) $\forall k, k' \neq 0$. Therefore, it must exist a Bogoliubov transformation linking the quasi-particle and the number-conserving operators, which, in general, are written as

$$c_k = e^{-i\alpha_k} \cosh \Lambda_k \delta \phi_k + e^{i\beta_k} \sinh \Lambda_k \delta \phi^\dagger_{-k}. \tag{125}$$

The transformation is defined through a set of functions Λ_k, constant in the stationary case, and the phases α_k and β_k, inheriting their time dependence from the atomic operators. These functions can be obtained from Equations (84) and (92):

$$\cosh \Lambda_k = \left(\frac{\omega_k + \left(\frac{k^2}{2m} + 2\lambda\rho_0 \right)}{4\omega_k} \right) \frac{\mathcal{N}_k}{\phi_0}. \tag{126}$$

If the coupling changes in time, the quasi-particle operators during the transient are defined knowing the solutions of the Klein–Gordon equation. With the Bogoliubov transformation of Equation (125), it is possible to find the quasi-particle vacuum-state $|\varnothing\rangle_{qp}$ in terms of the atomic degrees of freedom

$$\begin{aligned}|\varnothing\rangle_{qp} &= \prod_k \frac{e^{-\frac{1}{2} e^{i(\alpha_k+\beta_k)} \tanh \Lambda_k \delta\phi^\dagger_k \delta\phi^\dagger_{-k}}}{\cosh \Lambda_k} |\varnothing\rangle_a \\ &= \exp \sum_k \left(-\frac{1}{2} e^{i(\alpha_k+\beta_k)} \tanh \Lambda_k \delta\phi^\dagger_k \delta\phi^\dagger_{-k} - \ln \cosh \Lambda_k \right) |\varnothing\rangle_a , \end{aligned} \tag{127}$$

where $|\emptyset\rangle_a$ should be interpreted as the vacuum of excited atoms.

From Equation (127), it is clear that in the basis of atom occupation number, the quasi-particle vacuum is a complicated superposition of states with different number of atoms in the condensed 1-particle state (and a corresponding number of coupled excited atoms, in pairs of opposite momenta). Every correlation function is therefore dependent on the entanglement of this many-body atomic state.

This feature is enhanced by the dynamics, as can be observed from the scattering operator in Equation (119) relating early and late times. The scattering operator acts on atom pairs and the creation of quasi-particles affects the approximated vacuum differently depending on the number of atoms occupying the condensed 1-particle state. The creation of more pairs modifies further the superposition of the entangled atomic states depending on the total number of atoms and the initial number of excited atoms.

We can observe this from the action of the condensed state ladder operator, which does not commute with the the creation of coupled quasi-particles $c_k^\dagger c_{-k}^\dagger$, which is described by the combination of the operators $\delta\phi_k^\dagger \delta\phi_k$, $\delta\phi_k^\dagger \delta\phi_{-k}^\dagger$, and $\delta\phi_k \delta\phi_{-k}$. The ladder operator $a_0^\dagger$ commutes with the first, but not with the others:

$$\left(\delta\phi_k^\dagger \delta\phi_k\right) a_0^\dagger = a_0^\dagger \left(\delta\phi_k^\dagger \delta\phi_k\right), \tag{128}$$

$$\left(\delta\phi_k^\dagger \delta\phi_{-k}^\dagger\right)^n a_0^\dagger = a_0^\dagger \left(\delta\phi_k^\dagger \delta\phi_{-k}^\dagger\right)^n \left(\frac{N_0+1}{N_0+1-2n}\right)^{1/2}, \tag{129}$$

$$\left(\delta\phi_k \delta\phi_{-k}\right)^n a_0^\dagger = a_0^\dagger \left(\delta\phi_k \delta\phi_{-k}\right)^n \left(\frac{N_0+1}{N_0+1+2n}\right)^{1/2}. \tag{130}$$

The operators a_0 and $a_0^\dagger$ do not commute with the number-conserving atomic ladder operators, and therefore the creation of couples and the correlation functions, up to any order, will present corrections of order $1/N$ to the values that could be expected in the usual Bogoliubov description. Such corrections appear in correlation functions between quasi-particle operators and for correlations between quasi-particles and condensate atoms. This is equivalent to saying that a condensed state, which is generally not coherent, will present deviations from the expected correlation functions predicted by the Bogoliubov theory, due to both the interaction and the features of the initial state itself (through contributions coming from connected expectation values).

7.2. Entanglement Structure in Number-Conserving formalism

Within the Bogoliubov description discussed in Section 3, the mean-field approximation for the condensate is most adequate for states close to coherence, thus allowing a separate analysis for the mean-field. The field operator is split in the mean-field function $\langle\phi\rangle$ and the fluctuation operator $\delta\phi$, which is assumed not to affect the mean-field through backreaction. Therefore, the states in this picture can be written as

$$|\langle\phi\rangle\rangle_{mf} \otimes \left|\delta\phi, \delta\phi^\dagger\right\rangle_{a\,Bog}, \tag{131}$$

meaning that the state belongs to the product of two Hilbert spaces: the mean-field defined on one and the fluctuations on the other, with $\delta\phi$ and $\delta\phi^\dagger$ ladder operators acting only on the second. The Bogoliubov transformation from atom operators to quasi-particles allows to rewrite the state as shown in Equation (127). The transformation only affects its second part:

$$|\langle N\rangle\rangle_{mf} \otimes |\emptyset\rangle_{qp\,Bog} = |\langle N\rangle\rangle_{mf} \otimes \sum_{lr} a_{lr} |l, r\rangle_{a\,Bog}. \tag{132}$$

With such transformation, the condensed part of the state is kept separate from the superposition of coupled atoms (which here are denoted l and r for brevity) forming the excited part, a separation, that is maintained during the evolution in the Bogoliubov description. Also, the Bogoliubov transformation from early-times quasi-particles to late-times quasi-particles affects only the second part

$$|\langle N\rangle\rangle_{mf} \otimes \sum_{lr} a_{lr} |l,r\rangle_{a\,Bog} \Rightarrow |\langle N\rangle\rangle_{mf} \otimes \sum_{lr} a'_{lr} |l,r\rangle_{a\,Bog} . \tag{133}$$

In the number-conserving framework there is not such a splitting of the Fock space, and there is no separation between the two parts of the state. In this case, the best approximation for the quasi–particle vacuum is given by a superposition of coupled excitations of the atom operators, but the total number of atoms cannot be factored out:

$$|N;\varnothing\rangle_{qp} \approx \sum_{lr} a_{lr} |N-l-r,l,r\rangle_a . \tag{134}$$

The term in the RHS is a superposition of states with N total atoms, of which $N-l-r$ are in the condensed 1-particle state, and the others occupy excited atomic states and are coupled with each other analogously to the previous Equation (132) (the difference being the truncation of the sum, required for a sufficiently large number of excited atoms, implying a different normalization).

The evolution does not split the Hilbert space, and the final state will be a different superposition of atomic states:

$$\sum_{lr} a_{lr} |N,l,r\rangle_a \Rightarrow \sum_{lr} a'_{lr} \left(1 + \mathcal{O}\left(N^{-1}\right)\right) |N-l-r,l,r\rangle_a . \tag{135}$$

We remark that in the RHS the final state must include corrections of order $1/N$ with respect to the Bogoliubov prediction, due to the fully quantum behavior of the condensate ladder operators. These are small corrections, but we expect that the difference from the Bogoliubov prediction will be relevant when considering many-point correlation functions.

Moreover, these corrections remark the fact that states with different number of atoms in the condensate are transformed differently. If we consider a superposition of states of the type in Equation (134) with different total atom numbers so to reproduce the state in Equation (132), therefore replicating the splitting of the state, we would find that the evolution produces a final state with a different structure, because every state in the superposition evolves differently. Therefore, also assuming that the initial state could be written as

$$\sum_N \frac{e^{-N/2}}{\sqrt{N!}} |N;\varnothing\rangle_{qp} \approx |\langle N\rangle\rangle_{mf} \otimes |\varnothing\rangle_{qp\,Boq} , \tag{136}$$

anyway, the final state would unavoidably have different features:

$$\sum_N \frac{e^{-N/2}}{\sqrt{N!}} \sum_{lr} a'_{lr} \left(1 + \mathcal{O}\left(N^{-1}\right)\right) |N-l-r,l,r\rangle_a \neq |\langle N\rangle\rangle_{mf} \otimes \sum_{lr} a'_{lr} |l,r\rangle_{a\,Bog} . \tag{137}$$

We remark that our point is qualitative. Indeed, it is true that also in the weakly interacting limit the contribution coming from the interaction of Bogoliubov quasi-particles may be quantitatively larger than the $\mathcal{O}\left(N^{-1}\right)$ term in Equation (137). However, even if one treats the operator a_0 as a number disregarding its quantum nature, then one cannot have the above discussed entanglement. In that case, the Hilbert space does not have a sector associated to the condensed part and no correlation between the condensate and the quasi-particles is present. To have them one has to keep the quantum nature of a_0, and its contribution to the Hilbert space.

Alternatively, let us suppose to have an interacting theory of bosons for which no interactions between quasi-particles are present (as in principle one could devise and engineer similar models based on solvable interacting bosonic systems [33]). Even in that case one would have a qualitative difference (and the absence or presence of the entanglement structure here discussed) if one retains or not the quantum nature of a_0 and its contribution to the Hilbert space. Of course one could always argue that in principle the coupling between the quantum gravity and the matter degrees of freedom may be such to preserve the factorization of an initial state. This is certainly possible in principle, but it would require a surprisingly high degree of fine tuning at the level of the fundamental theory.

In conclusion, in the Bogoliubov description the state is split in two sectors, and the total density matrix is therefore a product of two contributions, of which the one relative to the mean-field can be traced away without affecting the other. The number-conserving picture shows instead that unavoidably the excited part of the system cannot be manipulated without affecting the condensate. Tracing away the quantum degrees of freedom of the condensate would imply a loss of information even without tracing away part of the couples created by analogous curved spacetime dynamics. In other words, when one considers the full Hilbert space and the full dynamics, the final state ρ_{fin} is obtained by an unitary evolution. But now, unlike the usual Bogoliubov treatment, one can trace out in ρ_{fin} the condensate degrees of freedom of the Hilbert space, an operation that we may denote by "$\text{Tr}_0[\ldots]$". So

$$\rho_{fin}^{reduced} = \text{Tr}_0[\rho_{fin}] \tag{138}$$

is not pure, as a consequence of the presence of the correlations. So one has $\text{Tr}[\rho_{fin}^2] = 1$, at variance with $\text{Tr}[(\rho_{fin}^{reduced})^2] \neq 1$. The entanglement between condensate and excited part is an unavoidable feature of the evolution of these states.

8. Discussion and Conclusions

The general aim of analogue gravity is to reproduce the phenomenology of quantum field theory on curved spacetime with laboratory-viable systems. In this framework, the geometry is given by a metric tensor assumed to be a classical tensor field without quantum degrees of freedom, implying that geometry and matter—the two elements of the system—are decoupled, i.e., the fields belong to distinct Hilbert spaces.

The usual formulation of analogue gravity in Bose–Einstein condensates reproduces this feature. In the analogy between the quasi-particle excitations on the condensate and those of scalar quantum fields in curved spacetime, the curvature is simulated by the effective acoustic metric derived from the classical condensate wave function. The condensate wave function itself does not belong to the Fock space of the excitations, instead it is a distinct classical function.

Moreover, analogue gravity with Bose–Einstein condensates is usually formulated assuming a coherent initial state, with a formally well-defined mean-field function identified with the condensate wave function. The excited part is described by operators obtained translating the atom field by the mean-field function and linearizing its dynamics; the quasi-particles studied in analogue gravity emerge from the resulting Bogoliubov–de Gennes equations. (Moreover, let us notice that the relation between several quantum gravity scenarios and analogue gravity in Bose–Einstein condensates appears to be even stronger than expected, as in many of these models a classical spacetime is recovered by considering an expectation value of the geometrical quantum degrees of freedom over a global coherent state the same way that the analogue metric is introduced by taking the expectation value of the field on a coherent ground state (see, e.g., the works by the authors of [34,35]). It is also interesting to note that within the AdS/CFT correspondence a deep connection between the analogue gravity system built from the hydrodynamics on the boundary and gravity in the bulk has emerged in recent work (see, e.g., the work by the authors of [36]). It would be interesting to extend the lessons of this work to these settings) The mean-field drives the evolution of the quasi-particles, which have a negligible backreaction on the condensate, and can be assumed to evolve independently, in accordance with the Gross–Pitaevskii equation. Neglecting the quantum nature of the operator creating particles in the condensate, one still has unitary dynamics occurring in the Hilbert space of noncondensed atoms (or, equivalently, of the quasi-particles).

However, one could still expect an information loss problem to arise whenever simulating an analogue black hole system entailing the complete loss of the ingoing Hawking partners, e.g., by having a flow with a region where the hydrodynamical description at the base of the standard analogue gravity formalism fails. The point is that the Bogoliubov theory is not exact as much as quantum field theory in curved spacetime is not a full description of quantum gravitational and matter degrees of freedom.

Within the number-conserving formalism, analogue gravity provides the possibility to develop a more complete description in which one is forced to retain the quantum nature of the operator creating particles in the condensate. While this is not per se a quantum gravity analogue (in the sense that it cannot reproduce the full dynamical equations of the quantum system), it does provide a proxy for monitoring the possible development of entanglement between gravitational and matter degrees of freedom.

It has already been conjectured in quantum gravity that degrees of freedom hidden from the classical spacetime description, but correlated to matter fields, are necessary to maintain unitarity in the global evolution and prevent the information loss [37]. To address the question of whether particle production induces entanglement between gravitation and matter degrees of freedom, we have carefully investigated the number-conserving formalism and studied the simpler process of cosmological particle production in analogue gravity, realized by varying the coupling constant from an initial value to a final one. We verified that one has a structure of quasi-particles, whose operators now depend on the operator a_0 destroying a particle in the natural orbital associated to the largest, macroscopic eigenvalue of the 1-point correlation functions.

We have shown that also in the number-conserving formalism one can define a unitary scattering operator, and thus the Bogoliubov transformation from early-times to late-times quasi-particles. The scattering operator provided in Equation (119) not only shows the nature of quasi-particle creation as a squeezing process of the initial quasi-particle vacuum, but also that the evolution process as a whole is unitary precisely, because it entangles the quasi-particles with the condensate atoms constituting the geometry over which the former propagate.

The correlation between the quasi-particles and the condensate atoms is a general feature, it is not realized just in a regime of high energies—analogous to the late stages of a black hole evaporation process or to sudden cosmological expansion—but it happens during all the evolution (Indeed, the transplanckian problem in black hole radiation may suggest that Hawking quanta might always probe the fundamental degrees of freedom of the underlying the geometry), albeit they are suppressed in the number of atoms, N, relevant for the system and are hence generally negligible. When describing the full Fock space, there is not unitarity breaking, and the purity of the state is preserved: it is not retrieved at late times nor is it spoiled in the transient of the evolution. Nonetheless, such a state after particle production will not factorise into the product of two states—a condensate (geometrical) and quasi-particle (matter) one—but, as we have seen, it will be necessarily an entangled state. This implies, as we have discussed at the end of the previous section, that an observer unable to access the condensate (geometrical) quantum degrees of freedom would define a reduced density matrix (obtained by tracing over the latter), which would no more be compatible with an unitary evolution.

In practice, in cases such as the cosmological particle creation, where the phenomenon happens on the whole spacetime, N is the (large) number of atoms in the whole condensate, and thus the correlations between the substratum and the quasi-particles are negligible. Therefore, the number-conserving formalism or the Bogoliubov one in this case may be practically equivalent. In the black hole case, a finite region of spacetime is associated to the particle creation, thus N is not only finite but decreases as a consequence of the evaporation making the correlators between geometry and Hawking quanta more and more non-negligible in the limit in which one simulates a black hole at late stages of its evaporation. This implies that tracing over the quantum geometry degrees of freedom could lead to non-negligible violation of unitary even for regular black hole geometries (i.e., for geometries without inner singularities, see, e.g., the works by the authors of [38–40]).

The Bogoliubov limit corresponds to taking the quantum degrees of freedom of the geometry as classical. This is not per se a unitarity violating operation, as it is equivalent to effectively recover the factorization of the above mentioned state. Indeed, the squeezing operator so recovered (which corresponds to the one describing particle creation on a classical spacetime) is unitarity preserving. However, the two descriptions are no longer practically equivalent when a region of quantum gravitational evolution is somehow simulated. In this case, having the possibility of tracking the quantum degrees of freedom underlying the background enables to describe the full evolution; whereas, in the analogue of quantum field theory in curved spacetime, a trace over the ingoing Hawking quanta is necessary with the usual problematic implications for the preservation of unitary evolution.

In the analogue gravity picture, the above alternatives would correspond to the fact that the number-conserving evolution can keep track of the establishment of correlations between the atoms and the quasi-particles that cannot be accounted for in the standard Bogoliubov framework. Hence, this analogy naturally leads to the conjecture that a full quantum gravitational description of a black hole creation and evaporation would leave not just a thermal bath on a Minkowski spacetime but rather a highly entangled state between gravitational and matter quantum degrees of freedom corresponding to the same classical geometry (With the possible exception of the enucleation of a disconnected baby universe which would lead to a sort of trivial information loss); a very complex state, but nonetheless a state that can be obtained from the initial one (for gravity and matter) via a unitary evolution.

In conclusion, the here presented investigation strongly suggests that the problems of unitarity breaking and information loss encountered in quantum field theory on curved spacetimes can only be addressed in a full quantum gravity description able to keep track of the correlations between quantum matter fields and geometrical quantum degrees of freedom developed via particle creation from the vacuum; these degrees of freedom are normally concealed by the assumption of a classical spacetime, but underlay it in any quantum gravity scenario.

Author Contributions: All the authors contributed equally to this work.

Funding: This research received no external funding.

Acknowledgments: Discussions with Renaud Parentani, Matt Visser, and Silke Weinfurtner are gratefully acknowledged.

Conflicts of Interest: The authors declare no conflicts of interest.

References

1. Unruh, W. Experimental black hole evaporation. *Phys. Rev. Lett.* **1981**, *46*, 1351–1353. [CrossRef]
2. Barceló, C.; Liberati, S.; Visser, M. Analogue gravity. *Living Rev. Rel.* **2005**, *8*, 12. [CrossRef] [PubMed]
3. Visser, M.; Barceló, C.; Liberati, S. Analog models of and for gravity. *Gen. Rel. Grav.* **2002**, *34*, 1719–1734. [CrossRef]
4. Girelli, F.; Liberati, S.; Sindoni, L. Gravitational dynamics in Bose Einstein condensates. *Phys. Rev.* **2008**, *D78*, 084013. [CrossRef]
5. Finazzi, S.; Liberati, S.; Sindoni, L. Cosmological Constant: A Lesson from Bose-Einstein Condensates. *Phys. Rev. Lett.* **2012**, *108*, 071101. [CrossRef] [PubMed]
6. Belenchia, A.; Liberati, S.; Mohd, A. Emergent gravitational dynamics in a relativistic Bose-Einstein condensate. *Phys. Rev.* **2014**, *D90*, 104015. [CrossRef]
7. Garay, L.; Anglin, J.; Cirac, J.; Zoller, P. Black holes in Bose-Einstein condensates. *Phys. Rev. Lett.* **2000**, *85*, 4643–4647. [CrossRef] [PubMed]
8. Garay, L.; Anglin, J.; Cirac, J.; Zoller, P. Sonic black holes in dilute Bose-Einstein condensates. *Phys. Rev.* **2001**, *A63*, 023611. [CrossRef]
9. Barceló, C.; Liberati, S.; Visser, M. Probing semiclassical analog gravity in Bose-Einstein condensates with widely tunable interactions. *Phys. Rev.* **2003**, *A68*, 053613. [CrossRef]
10. Lahav, O.; Itah, A.; Blumkin, A.; Gordon, C.; Steinhauer, J. Realization of a sonic black hole analogue in a Bose-Einstein condensate. *Phys. Rev. Lett.* **2010**, *105*, 240401. [CrossRef]

11. Steinhauer, J. Measuring the entanglement of analogue Hawking radiation by the density-density correlation function. *Phys. Rev.* **2015**, *D92*, 024043. [CrossRef]
12. Steinhauer, J. Observation of quantum Hawking radiation and its entanglement in an analogue black hole. *Nat. Phys.* **2016**, *12*, 959. [CrossRef]
13. Muñoz de Nova, J.R.; Golubkov, K.; Kolobov, V.I.; Steinhauer, J. Observation of thermal Hawking radiation and its temperature in an analogue black hole. *Nature* **2019**, *569*, 688–691. [CrossRef]
14. Giddings, S.B. Comments on information loss and remnants. *Phys. Rev.* **1994**, *D49*, 4078–4088. [CrossRef] [PubMed]
15. Giddings, S.B. Constraints on black hole remnants. *Phys. Rev.* **1994**, *D49*, 947–957. [CrossRef] [PubMed]
16. Susskind, L. Trouble for remnants. *arXiv* **1995**, arXiv:hep-th/9501106.
17. Chen, P.; Ong, Y.C.; Yeom, D. Black Hole Remnants and the Information Loss Paradox. *Phys. Rept.* **2015**, *603*, 1–45. [CrossRef]
18. Hossenfelder, S.; Smolin, L. Conservative solutions to the black hole information problem. *Phys. Rev.* **2010**, *D81*, 064009. [CrossRef]
19. Almheiri, A.; Marolf, D.; Polchinski, J.; Sully, J. Black holes: Complementarity or firewalls? *J. High Energy Phys.* **2013**, *2013*, 62. [CrossRef]
20. Pethick, C.J.; Smith, H. *Bose-Einstein Condensation in Dilute Gases*, 2nd ed.; Cambridge Cambridge University Press Press: Cambridge, UK, 2008.
21. Zagrebnov, V.A. The Bogoliubov model of weakly imperfect Bose gas. *Phys. Rep.* **2001**, *350*, 291–434. [CrossRef]
22. Leggett, A. *Quantum Liquids: Bose Condensation and Cooper Pairing in Condensed-Matter Systems*; Oxford University Press: Oxford, UK, 2008. [CrossRef]
23. Penrose, O.; Onsager, L. Bose-Einstein Condensation and Liquid Helium. *Phys. Rev.* **1956**, *104*, 576–584. [CrossRef]
24. Pitaevskii, L.P.; Stringari, S. *Bose-Einstein Condensation*; Clarendon Press: Oxford, UK, 2003.
25. Lieb, E.H.; Seiringer, R.; Solovej, J.P.; Yngvason, J. *The Mathematics of the Bose Gas and its Condensation*; Oberwolfach Seminar Series; Birkhäuser: Basel, Switzerland, 2006.
26. Proukakis, N.P.; Jackson, B. Finite-temperature models of Bose–Einstein condensation. *J. Phys. B* **2008**, *41*, 203002. [CrossRef]
27. Jain, P.; Weinfurtner, S.; Visser, M.; Gardiner, C.W. Analog model of a Friedmann-Robertson-Walker universe in Bose-Einstein condensates: Application of the classical field method. *Phys. Rev. A* **2007**, *76*, 033616. [CrossRef]
28. Weinfurtner, S.; Jain, P.; Visser, M.; Gardiner, C.W. Cosmological particle production in emergent rainbow spacetimes. *Class. Quantum Gravity* **2009**, *26*, 065012. [CrossRef]
29. Carusotto, I.; Balbinot, R.; Fabbri, A.; Recati, A. Density correlations and analog dynamical Casimir emission of Bogoliubov phonons in modulated atomic Bose-Einstein condensates. *Eur. Phys. J. D* **2010**, *56*, 391–404. [CrossRef]
30. Jaskula, J.C.; Partridge, G.B.; Bonneau, M.; Lopes, R.; Ruaudel, J.; Boiron, D.; Westbrook, C.I. Acoustic Analog to the Dynamical Casimir Effect in a Bose-Einstein Condensate. *Phys. Rev. Lett.* **2012**, *109*, 220401. [CrossRef] [PubMed]
31. Sotiriadis, S.; Calabrese, P. Validity of the GGE for quantum quenches from interacting to noninteracting models. *J. Stat. Mech. Theory Exp.* **2014**, *2014*, P07024. [CrossRef]
32. Birrell, N.D.; Davies, P.C.W. *Quantum Fields in Curved Space*; Cambridge Cambridge University Press Press: Cambridge, UK, 1984. [CrossRef]
33. Richardson, R.W. Exactly Solvable Many? Boson Model. *J. Math. Phys.* **1968**, *9*, 1327–1343. [CrossRef]
34. Alesci, E.; Bahrami, S.; Pranzetti, D. Quantum Gravity Predictions for Black Hole Interior Geometry. *arXiv* **2019**, arXiv:gr-qc/1904.12412.
35. Gielen, S.; Oriti, D.; Sindoni, L. Homogeneous cosmologies as group field theory condensates. *JHEP* **2014**, *6*, 13. [CrossRef]
36. Ge, X.H.; Sun, J.R.; Tian, Y.; Wu, X.N.; Zhang, Y.L. Holographic interpretation of acoustic black holes. *Phys. Rev. D* **2015**, *92*, 084052. [CrossRef]
37. Perez, A. No firewalls in quantum gravity: The role of discreteness of quantum geometry in resolving the information loss paradox. *Class. Quant. Grav.* **2015**, *32*, 084001. [CrossRef]

38. Frolov, V.P.; Vilkovisky, G.A. Spherically Symmetric Collapse in Quantum Gravity. *Phys. Lett.* **1981**, *106*, 307–313. [CrossRef]
39. Hayward, S.A. Formation and evaporation of regular black holes. *Phys. Rev. Lett.* **2006**, *96*, 031103. [CrossRef] [PubMed]
40. Carballo-Rubio, R.; Di Filippo, F.; Liberati, S.; Pacilio, C.; Visser, M. On the viability of regular black holes. *JHEP* **2018**, *7*, 23. [CrossRef]

© 2019 by the authors. Licensee MDPI, Basel, Switzerland. This article is an open access article distributed under the terms and conditions of the Creative Commons Attribution (CC BY) license (http://creativecommons.org/licenses/by/4.0/).

Article

A Physically-Motivated Quantisation of the Electromagnetic Field on Curved Spacetimes

Ben Maybee [1], Daniel Hodgson [2], Almut Beige [2] and Robert Purdy [2],*

[1] Higgs Centre for Theoretical Physics, School of Physics and Astronomy, The University of Edinburgh, Edinburgh EH9 3JZ, UK
[2] The School of Physics and Astronomy, University of Leeds, Leeds LS2 9JT, UK
* Correspondence: r.purdy@leeds.ac.uk

Received: 6 July 2019; Accepted: 26 August 2019; Published: 30 August 2019

Abstract: Recently, Bennett et al. (Eur. J. Phys. 37:014001, 2016) presented a physically-motivated and explicitly gauge-independent scheme for the quantisation of the electromagnetic field in flat Minkowski space. In this paper we generalise this field quantisation scheme to curved spacetimes. Working within the standard assumptions of quantum field theory and only postulating the physicality of the photon, we derive the Hamiltonian, $\hat{H}$, and the electric and magnetic field observables, $\hat{\mathbf{E}}$ and $\hat{\mathbf{B}}$, respectively, without having to invoke a specific gauge. As an example, we quantise the electromagnetic field in the spacetime of an accelerated Minkowski observer, Rindler space, and demonstrate consistency with other field quantisation schemes by reproducing the Unruh effect.

Keywords: quantum electrodynamics; relativistic quantum information

1. Introduction

For many theorists the question "what is a photon?" remains highly nontrivial [1]. It is in principle possible to uniquely define single photons in free space [2]; however, the various roles that photons play in light–matter interactions [3], the presence of boundary conditions in experimental scenarios [4,5] and our ability to arbitrarily shape single photons [6] all lead to a multitude of possible additional definitions. Yet this does not stop us from utilising single photons for tasks in quantum information processing, especially for quantum cryptography, quantum computing, and quantum metrology [7]. In recent decades, it has become possible to produce single photons on demand [8], to transmit them over 100 kilometres through Earth's atmosphere [9] and to detect them with very high efficiencies [10]. Moreover, single photons have been an essential ingredient in experiments probing the foundations of quantum physics, such as entanglement and locality [11,12].

Recently, relativistic quantum information has received a lot of attention in the literature. Pioneering experiments verify the possibility of quantum communication channels between Earth's surface and space [13] and have transmitted photons between the Earth and low-orbit satellites [14], while quantum information protocols are beginning to extend their scope towards the relativistic arena [15–21]. The effects of gravity on satellite-based quantum communication schemes, entanglement experiments and quantum teleportation have already been shown to produce potentially observable effects [22–25]. Noninertial motion strongly affects quantum information protocols and quantum optics set-ups [26–30], with the mere propagation and detection of photons in such frames being highly nontrivial [31–34].

Motivated by these recent developments, this paper generalises a physically-motivated quantisation scheme of the electromagnetic field in flat Minkowski space [35] to curved space times. Our approach aims to obtain the basic tools for analysing and designing relativistic quantum information experiments in a more direct way than alternative derivations, and without having to invoke a specific gauge. Working within the standard assumptions of quantum field theory and only postulating the physicality of the photon, we derive the Hamiltonian, $\hat{H}$, and the observables, $\hat{\mathbf{E}}$ and $\hat{\mathbf{B}}$, of the electromagnetic field. Retaining gauge-independence is important when modelling the interaction of the electromagnetic field with another quantum system, like an atom. In this case, different choices of gauge correspond to different subsystem decompositions, thereby affecting our notion of what is 'atom' and what is 'field' [36,37]. Composite quantum systems can be decomposed into subsystems in many different ways. Choosing an unphysical decomposition can result in the prediction of spurious effects when analysing the dynamics of one subsystem while tracing out the degrees of freedom of the other [38]. Hence it is important to first formulate quantum electrodynamics in an entirely arbitrary gauge, as this allows us to subsequently fix the gauge when needed. This work does not seek to quantise the gravitational field. Instead, we follow the standard approach of quantum field theory in curved spacetime. This is a first approximation to understanding gravitational effects on quantum fields [39,40], which neglects the back-reaction of those fields on the spacetime geometry, treating the spacetime as a fixed background.

The direct canonical quantisation of the electromagnetic field in terms of the (real) gauge independent electric and magnetic fields, $\mathbf{E}$ and $\mathbf{B}$, is not possible, since these do not offer a complete set of canonical variables [41–45]. As an alternative, Bennett et al. [35] suggested to use the physicality of the photon as the starting point when quantising the electromagnetic field. Assuming that the electromagnetic field is made up of photons and identifying their relevant degrees of freedom, like frequencies and polarisations, results in a harmonic oscillator Hamiltonian $\hat{H}$ for the electromagnetic field. Using this Hamiltonian and demanding consistency of the dynamics of expectation values with classical electrodynamics, especially with Maxwell's equations, is sufficient to then obtain expressions for $\hat{\mathbf{E}}$ and $\hat{\mathbf{B}}$ without having to invoke vector potentials and without having to choose a specific gauge. Generalising the work by the authors of [35] from flat Minkowski space to curved space times, we obtain field observables which could be used, for example, to model photonics experiments in curved spacetimes in a similar fashion to how quantum optics typically models experiments in Minkowski space [5,36,46].

Additional problems with our understanding of photons (indeed all particles) arise when we consider quantum fields in gravitationally bound systems [7]. General relativity can be viewed as describing gravitation as the consequence of interactions between matter and the curvature of a Lorentzian (mixed signature) spacetime with metric $g_{\mu\nu}$ [47,48]. Locally, however, any spacetime appears flat, by which we mean

$$g_{\mu\nu}(p) \cong \eta_{\mu\nu} \equiv \mathrm{diag}(+1, -1, -1, -1), \tag{1}$$

the familiar special relativistic invariant line-element of Minkowski space. For the Earth's surface, where gravity is (nearly) uniform, this limit can be taken everywhere, and spacetime curvature can be neglected. Spacetimes in relativity have no preferred coordinate frame, so physical laws must satisfy the principle of covariance and be coordinate independent and invariant under coordinate transformations [49]. Indeed, it has been demonstrated that, while the form of the Hamiltonian may change under general coordinate transformations, physically measurable predictions do not [50].

Quantum field theory in curved spacetime is the standard approach used to study the behaviour of quantum fields in this setting. As aforementioned, this is a first approximation to quantum gravity, in which the gravitational field is treated classically and back-reactions on the spacetime geometry are neglected [39,40]. Intuitively this is what is meant by a static spacetime, where the time derivative of the metric is zero. This approximation holds on typical astrophysical length and energy scales and is thus well-suited for dealing with most physical situations [51]. How to generalise field quantisation to curved spaces is very well established, and the theory has produced several major discoveries, like the prediction that the particle states seen by a given observer depend on the geometry of their spacetime [52–54]. For example, the vacuum state of one observer does not necessarily coincide with the vacuum state of an observer in an alternative reference frame. This surprising result even arises in flat Minkowski space, where the Fulling–Davies–Unruh effect predicts that an observer with constant acceleration sees the Minkowski vacuum as a thermal state with temperature proportional to their acceleration [55–59].

To make quantum field theory in curved spacetimes more accessible to quantum opticians, and to obtain more insight into the aforementioned effects and their experimental ramifications, this paper considers static, 4-dimensional Lorentzian spacetimes. Our starting point for the derivation of the field observables $\hat{H}$, $\hat{\mathbf{E}}$ and $\hat{\mathbf{B}}$ is the assumption that the detectors belonging to a moving observer see photons. These are the energy quanta of the electromagnetic field in curved space times. To demonstrate the consistency of our approach with other field quantisation schemes, we consider the explicit case of an accelerated Minkowski observer, who is said to reside in a Rindler spacetime [60–64], and reproduce Unruh's predictions [55–59].

This paper is divided into five sections. In Section 2, we provide a summary of the gauge-independent quantisation scheme by Bennett et al. [35] which applies in the case of flat spacetime. In Section 3, we discuss what modifications must be made to classical electrodynamics when moving to the more general setting of a stationary curved spacetime. We then show that similar modifications allow for the gauge-independent quantisation scheme of Section 2 to be applied in this more general setting. In Section 4, we apply our results to the specific case of a uniformly accelerating reference frame and have a closer look at the Unruh effect. Finally, we draw our conclusions in Section 5. For simplicity, we work in natural units $\hbar = c = 1$ throughout.

2. Gauge-Independent Quantisation of the Electromagnetic Field

In this section, we review the gauge dependence inherent in the electromagnetic field and contrast standard, more mathematically-motivated quantisation procedures with the gauge-independent method of Bennett et al. [35].

2.1. Classical Electrodynamics

Under coordinate transformations, the electric and magnetic fields transform as the components of an antisymmetric 2-form, the field strength tensor

$$F_{\mu\nu} = \begin{pmatrix} 0 & E^1 & E^2 & E^3 \\ -E^1 & 0 & -B^3 & B^2 \\ -E^2 & B^3 & 0 & -B^1 \\ -E^3 & -B^2 & B^1 & 0 \end{pmatrix}. \tag{2}$$

The field strength is defined in terms of the 4-vector potential by

$$F_{\mu\nu} = \partial_\mu A_\nu - \partial_\nu A_\mu. \tag{3}$$

We can obtain the equations of motion by applying the Euler–Lagrange equations to the Lagrangian density

$$\mathcal{L} = -\frac{1}{4} F_{\mu\nu} F^{\mu\nu} = \frac{1}{2} \left(\mathbf{E}^2 - \mathbf{B}^2 \right), \tag{4}$$

which gives the Maxwell equation

$$\partial_\mu F^{\mu\nu} = 0. \tag{5}$$

The field strength tensor also satisfies the Bianchi identity,

$$\partial_{[\sigma} F_{\mu\nu]} \equiv \frac{1}{3} \left(\partial_\sigma F_{\mu\nu} + \partial_\mu F_{\nu\sigma} + \partial_\nu F_{\sigma\mu} \right) = 0, \tag{6}$$

and together, Equations (5) and (6) can be used to obtain the standard Maxwell equations expressed in terms of the magnetic and electric field strengths, **E** and **B**, respectively,

$$\begin{aligned} \text{div}\,(\mathbf{E}) &= 0, & \text{curl}\,(\mathbf{B}) &= \dot{\mathbf{E}}, \\ \text{div}\,(\mathbf{B}) &= 0, & \text{curl}\,(\mathbf{E}) &= -\dot{\mathbf{B}}. \end{aligned} \qquad (7)$$

The solutions to these equations are transverse plane waves with orthogonal electric and magnetic field components with two distinct, physical polarisations propagating through Minkowski space, $\mathbb{M}$, at a speed $c = 1$.

2.2. Gauge Dependence in Electromagnetic Field Quantisation

The most commonly used methods for quantising fields are the traditional canonical and modern path-integral approaches. When applied to electromagnetism, these have to be modified due to the gauge freedom of the theory. For example, in the canonical approach, standard commutation relations cannot be satisfied. One can get around this by either breaking Lorentz invariance in intermediate steps of calculations, or by considering excess degrees of freedom with negative norms that do not contribute physically [37]. Standard path integral quantisation fails for electromagnetism because the resultant propagator is divergent. The Fadeev–Popov procedure rectifies this by implementing a gauge-fixing condition [65]. This method also gives additional terms from nonphysical contributions in the form of Fadeev–Popov ghosts. Such terms can be ignored for free fields in Minkowski space as they only appear in loop diagrams, but in curved spacetimes this is not the case [51]. While physical quantities remain gauge-invariant under both approaches to quantisation, nondirectly observable quantities can become gauge-dependent.

This can result in conceptual problems when modelling composite quantum systems, like the ones that are of interest to those working in relativistic quantum information, quantum optics and condensed matter. Suppose H denotes the total Hamiltonian of a composite quantum system. Then one can show that any Hamiltonian H' of the form

$$H' = U^\dagger H U, \qquad (8)$$

where U denotes a unitary operator, has the same energy eigenvalues as H. Both Hamiltonians H and H' are unitarily equivalent and can be used interchangeably. However, the dynamics of subsystem observables O can depend on the concrete choice of U, since $O' = U^\dagger O U$ and O are in general not the same. For example, atom–field interactions depend on the gauge-dependent vector potential **A** for most subsystem decompositions [36,37]. Hence it is important here to formulate quantum electrodynamics in an entirely arbitrary gauge and to maintain ambiguity as long as possible, thereby retaining the ability to later choose a gauge which does not result in the prediction of spurious effects [38].

2.3. Physically-Motivated Gauge-Independent Method

In contrast to this, the electromagnetic field quantisation scheme presented in the work by the authors of [35] relies upon two primary experimentally derived assumptions. Firstly, the electric and the magnetic field expectation values follow Maxwell's equations, and, secondly, the field is composed of photons of energy $\hbar \omega_\mathbf{k}$, or $\omega_\mathbf{k}$ in natural units. Whereas, in standard canonical quantisation, the electromagnetic field's photon construction is a derived result, for the method of [35] it is an initial premise. This is physically acceptable, since photons are experimentally detectable entities [7,10]. The motivation for the scheme [35] comes from the observation that one observes discrete clicks when measuring a very weak electromagnetic field. An experimental definition of photons is that these are electromagnetic field excitations with the property that their integer numbers can be individually detected, given a perfect detector [10].

Hence the Fock space for this gauge-independent approach is spanned by states of the form

$$\bigotimes_{\lambda=1,2} \bigotimes_{k^1=-\infty}^{\infty} \bigotimes_{k^2=-\infty}^{\infty} \bigotimes_{k^3=-\infty}^{\infty} |n_{\mathbf{k}\lambda}\rangle , \qquad (9)$$

where $n_{\mathbf{k}\lambda}$ is the number of excitations of a mode with wave-vector $\mathbf{k}$ and physical, transverse polarisation state λ. Since it is an experimental observation that photons of frequency $\omega_{\mathbf{k}} = |\mathbf{k}|$ have energy $\omega_{\mathbf{k}}$ in natural units, the Hamiltonian $\hat{H}$ for such a Fock space must satisfy

$$\hat{H} |n_{\mathbf{k}\lambda}\rangle = (\omega_{\mathbf{k}} n_{\mathbf{k}\lambda} + H_0) |n_{\mathbf{k}\lambda}\rangle , \qquad (10)$$

where H_0 is the vacuum or zero-point energy and $n_{\mathbf{k}\lambda}$ is an integer [35]. An infinite set of evenly spaced energy levels, as is present here, has been proven to be unique to the simple harmonic oscillator [66]. Hence this Hamiltonian must take the form [5]

$$\hat{H} = \sum_{\lambda=1,2} \int d^3k \left(\omega_{\mathbf{k}} \hat{a}^\dagger_{\mathbf{k}\lambda} \hat{a}_{\mathbf{k}\lambda} + H_0 \right) , \qquad (11)$$

where the $\hat{a}_{\mathbf{k}\lambda}, \hat{a}^\dagger_{\mathbf{k}\lambda}$ are a set of independent ladder operators for each $(\mathbf{k}, \lambda)$ mode, obeying the canonical commutation relations

$$[\hat{a}_{\mathbf{k}\lambda}, \hat{a}_{\mathbf{k}'\lambda'}] = 0, \quad [\hat{a}^\dagger_{\mathbf{k}\lambda}, \hat{a}^\dagger_{\mathbf{k}'\lambda'}] = 0, \quad [\hat{a}_{\mathbf{k}\lambda}, \hat{a}^\dagger_{\mathbf{k}'\lambda'}] = \delta_{\lambda\lambda'} \delta^3(\mathbf{k} - \mathbf{k}') . \qquad (12)$$

Since the classical energy density is quadratic in the electric and magnetic fields, while the above Hamiltonian is quadratic in the ladder operators, the field operators must be linear superpositions of creation and annihilation operators [35]. By further demanding that the fields' expectation values satisfy Maxwell's equations, consistency with the Heisenberg equation of motion,

$$\frac{\partial}{\partial t} \hat{O} = -i[\hat{O}, \hat{H}] , \qquad (13)$$

allows the coefficients of these superpositions to be deduced, and the (Heisenberg) field operators can be shown to be of the form [35]

$$\begin{aligned}
\hat{\mathbf{E}}(\mathbf{x},t) &= i \sum_{\lambda=1,2} \int d^3k \sqrt{\frac{\omega_{\mathbf{k}}}{16\pi^3}} [e^{i(\mathbf{k}\cdot\mathbf{x}-\omega_{\mathbf{k}}t)} \hat{a}_{\mathbf{k}\lambda} + \text{H.c.}] \hat{\mathbf{e}}_\lambda , \\
\hat{\mathbf{B}}(\mathbf{x},t) &= -i \sum_{\lambda=1,2} \int d^3k \sqrt{\frac{\omega_{\mathbf{k}}}{16\pi^3}} [e^{i(\mathbf{k}\cdot\mathbf{x}-\omega_{\mathbf{k}}t)} \hat{a}_{\mathbf{k}\lambda} + \text{H.c.}] (\hat{\mathbf{k}} \times \hat{\mathbf{e}}_\lambda) ,
\end{aligned} \qquad (14)$$

where $\hat{\mathbf{e}}_\lambda$ is a unit polarisation vector orthogonal to the direction of propagation, with $\hat{\mathbf{e}}_1 \cdot \hat{\mathbf{e}}_2 = \hat{\mathbf{e}}_1 \cdot \mathbf{k} = \hat{\mathbf{e}}_2 \cdot \mathbf{k} = 0$. This is also consistent with the Hamiltonian being a direct operator-valued promotion of its classical form

$$\hat{H}(t) = \frac{1}{2} \int d^3x \left[\hat{\mathbf{E}}^2(\mathbf{x},t) + \hat{\mathbf{B}}^2(\mathbf{x},t) \right] . \qquad (15)$$

Comparing Equations (11) and (15) allows us to determine the zero point energy H_0 in Minkowski space, which coincides with the energy expectation value of the vacuum state $|0\rangle$ of the electromagnetic field. In quantum optics, Equations (11) and (14) often serve as the starting point for further investigations [5,36,46].

Note that a quantisation scheme in a similar spirit to the work by the authors of [35] can be found in the work by the authors of [67], which also uses the Maxwell and Heisenberg equations to directly quantise the physical field operators. The attraction of such a scheme is in the lack of reliance on the gauge-dependent electromagnetic potentials, instead directly quantising the gauge-invariant electric and magnetic fields, the benefits of which for quantum optics were discussed in the preceding section.

3. Gauge-Independent Quantisation of the Electromagnetic Field in Curved Spacetimes

Many aspects of the quantisation method of Bennett et al. [35] are explicitly noncovariant, and hence unsuitable for general curved spacetimes. Here we lift the scheme onto static spacetimes, maintaining the original global structure and approach.

3.1. Classical Electrodynamics in Curved Space

To begin, consider electromagnetism on stationary spacetimes in general relativity, which are differentiable manifolds with a metric structure $g_{\mu\nu}$. By stationary we mean $\partial_0 g_{\mu\nu} = 0$. For any theory, the standard approach is to follow the minimal-coupling procedure [48,51],

$$
\begin{aligned}
\eta_{\mu\nu} &\to g_{\mu\nu}, \\
\partial_\mu &\to \nabla_\mu, \\
\int d^4x &\to \int d^4x \sqrt{|g|},
\end{aligned}
\qquad (16)
$$

where $g = \det(g_{\mu\nu})$ and ∇_μ is the covariant derivative associated with the metric (Levi-Civita) connection. Since electric and magnetic fields can be expressed in a covariant form through the field strength tensor, it is simple to generalise to curved space by just applying this procedure. Firstly, the derivatives of the four-vector potential generalise to

$$
\begin{aligned}
\nabla_\nu A_\mu &= \partial_\nu A_\mu - \Gamma^\rho_{\mu\nu} A_\rho, \\
\nabla_\nu A^\mu &= \partial_\nu A^\mu + \Gamma^\mu_{\rho\nu} A^\rho,
\end{aligned}
\qquad (17)
$$

where $\Gamma^\mu_{\nu\rho}$ are the standard symmetric Christoffel symbols. The field strength tensor and the Bianchi identity remain unchanged by these derivatives, as their explicit antisymmetry cancels all the Christoffel symbols. Thus Equations (3) and (6) still hold in curved spacetimes. The only modification we need to make is to the (free-space) inhomogeneous Maxwell equation. Applying the minimal-coupling procedure to Equation (5) gives

$$
\nabla_\mu F^{\mu\nu} = 0, \qquad (18)
$$

which on stationary spacetimes can be written as [61]

$$
\nabla_\mu F^{\mu\nu} = \frac{1}{\sqrt{|g|}} \partial_\mu \left(\sqrt{|g|} F^{\mu\nu} \right) = 0, \qquad (19)
$$

as may be obtained from a Lagrangian density $\mathcal{L} = -\frac{1}{4}\sqrt{|g|} F_{\mu\nu} F^{\mu\nu}$. To obtain the modified Maxwell equations for the electric and magnetic field strengths, one may now simply extract the relevant terms from the covariant form given above, working in a particular coordinate system [60]. For the resulting wave equations, as with any wave equation on a curved spacetime, obtaining a general solution is a highly nontrivial task [49]. However, on simple spacetimes such as we will consider later, it is possible to obtain analytic solutions.

3.2. Particles in Curved Spacetimes

To quantise the electromagnetic field in the manner of [35] our starting point must be to write down an appropriate Fock space for experimentally observable photon states. On curved spacetimes this is complicated by the lack of a consistent frame-independent basis for such a space. To see why, consider that to introduce particle states in quantum field theory, we must first write the solutions to a momentum–space wave equation as a superposition of orthonormal field modes, which are split into positive and negative frequency modes (f_i, f_i^*). In order for us to do this, the spacetime must have a timelike symmetry. Symmetries of spacetimes are generated by Killing vectors, V, which satisfy

$$\nabla_\mu V_\nu + \nabla_\nu V_\mu = 0. \tag{20}$$

If V_μ is, in addition, timelike at asymptotic infinity then it defines a timelike Killing vector K_μ. The presence of such a vector defines a stationary spacetime, in which there always exists a coordinate frame such that $\partial_t g_{\mu\nu} = 0$, where $x^0 = t$ in this coordinate set is the Killing time. If, in addition, K^μ is always orthogonal to a family of spacelike hypersurfaces then the spacetime is said to be static, and in addition we have $g_{ti} = 0$. Conceptually, the spacetime background is fixed but fields can propagate and interact. Particle states can only be canonically introduced with frequency splitting. Hence, to define particles in a curved spacetime there must be a timelike Killing vector [53].

Canonical field quantisation morphs the field into an operator acting on a Fock space of particle states, promoting the coefficients of the positive frequency modes to annihilation operators and those of negative frequency modes to creation operators [33,40]. General field states are therefore critically dependent on the frequency splitting of the modes, which itself depends on the background geometry of the spacetime [52]. In general, we define positive and negative frequency modes f_{ω_k} of frequency ω_k with respect to the timelike Killing vector K^μ, by using the definition

$$\pounds_K f_{\omega_k} = \begin{cases} -i\omega_k f_{\omega_k} & \text{positive frequency} \\ i\omega_k f_{\omega_k} & \text{negative frequency} \end{cases}, \tag{21}$$

where $\pounds_K$ is the coordinate-invariant Lie derivative along K^μ, which, in this case, is given by $K^\mu \partial_\mu$. However, a particle detector reacts to states of positive frequency with respect to its own proper time τ, not the killing time [55]. For a timelike observer with worldline x^μ on a (not necessarily stationary) spacetime, the proper time is defined by the metric $g_{\mu\nu}$ infinitesimally as

$$d\tau = \sqrt{g_{\mu\nu} dx^\mu dx^\nu}. \tag{22}$$

A given detector with proper time τ has positive frequency modes g_{ω_k} satisfying

$$\frac{dx^\mu}{d\tau} \nabla_\mu g_{\omega_k} = -i\omega_k g_{\omega_k}, \tag{23}$$

and they will, generally, only cover part of the spacetime. To consistently approach quantisation we need these detector modes to relate to the set f_{ω_k} defined with respect to the timelike Killing vector. Fortunately, the set of modes f_{ω_k} forms a natural basis for the detector's Fock space if the proper time τ is proportional to the Killing time t. This occurs if the future-directed timelike Killing vector is tangent to the detector's trajectory [48,55].

Even with a timelike Killing vector and its associated symmetry, solving a given wave equation and hence obtaining mode solutions can still be highly nontrivial [49]. Considering a static spacetime greatly simplifies this as the d'Alembertian operator, $\Box = \nabla_\mu \nabla^\mu$, that appears in the general wave equation can be separated into pure spatial and temporal derivatives, allowing us to easily write separable mode solutions [48,52]

$$f_{\omega_k}(x) = e^{-i\omega_k t} \Sigma_{\omega_k}(\mathbf{x}). \tag{24}$$

These modes are then positive frequency in the above sense, and conjugate modes $f_{\omega_k}^*$ are negative frequency. The set $(f_{\omega_k}, f_{\omega_k}^*)$ then forms a complete basis of solutions for the wave equation and provides a suitable basis for particle detectors.

However, when two distinct inertial particle detectors follow different geodesic paths in the spacetime, each will have its own unique proper time, determined by its motion and the local geometry. But this proper time is what we have used in Equation (23) to define the basis modes associated with a given particle Fock space associated with a particle detector. Thus, the detectors will define the particle states they observe in different manners, and will not agree on a natural set of basis modes [52,53,68]. This has no counterpart in inertial Minkowski space, where there is a global Poincaré symmetry, but will be unavoidable in our scheme.

3.3. Covariant and Gauge-Independent Electromagnetic Field Quantisation Scheme

Accommodating for the above considerations allows the physically motivated scheme [35] to be covariantly generalised to static curved spacetimes.

3.3.1. Hilbert Space

Since for static spacetimes there exists a global timelike Killing vector we can define positive and negative frequency modes and thus introduce a well-defined particle Fock space. Again we assume the existence of photons on the considered spacetime. As travelling waves on the spacetime, these photons are again characterised by their physical, transverse polarisation λ and wave-vector $\mathbf{k}$ [69]. Taking these as labels for general states yields again the states in Equation (9) as the basis states of the quantised field. Physical energy eigenstates have integer values of $n_{\mathbf{k}\lambda}$ and are associated with energy $\omega_{\mathbf{k}}$. Thus the field Hamiltonian must again satisfy Equation (10), allowing it to be written in terms of independent ladder operators [66]. In the following, we denote these by $\hat{b}_{\mathbf{k}\lambda}$ and assume that they satisfy the equal time canonical commutation relations

$$[\hat{b}_{\mathbf{k}\lambda}, \hat{b}_{\mathbf{k}'\lambda'}] = 0, \quad [\hat{b}^\dagger_{\mathbf{k}\lambda}, \hat{b}^\dagger_{\mathbf{k}'\lambda'}] = 0, \quad [\hat{b}_{\mathbf{k}\lambda}, \hat{b}^\dagger_{\mathbf{k}'\lambda'}] = \delta_{\lambda\lambda'}\delta^3(\mathbf{k} - \mathbf{k}'). \tag{25}$$

Importantly, the $\hat{b}_{\mathbf{k}\lambda}$ generate a distinct Fock space from that of the ladder operators utilised in the Minkowskian case.

3.3.2. Hamiltonian

To write down the full field or classical Hamiltonian requires some care, as a Hamiltonian is a component of the energy–momentum tensor

$$T_{\mu\nu} = -\frac{2}{\sqrt{|g|}}\frac{\delta S_{\text{matter}}}{\delta g^{\mu\nu}}, \tag{26}$$

where S_{matter} is the action determining the matter content on the spacetime. As a component of a tensor, the Hamiltonian itself is not invariant under general coordinate transformations. On stationary spacetimes a conserved energy equal to the Hamiltonian can be introduced through the timelike Killing current

$$J^\mu = K_\nu T^{\mu\nu}, \tag{27}$$

which satisfies the continuity equation $\nabla_\mu J^\mu = 0$. Stokes's theorem can then be used to integrate over a spacelike hypersurface Σ in three dimensions, giving

$$H = \int_\Sigma d^3x\, \sqrt{|\gamma|} n_\mu J^\mu, \tag{28}$$

where $\gamma = \det(\gamma_{ij})$ with γ_{ij} being the induced metric on Σ and n^μ being the timelike unit normal vector to Σ. On stationary spacetimes the result of this integral is the same for all hypersurfaces Σ [48,70]. For the electromagnetic field, the variation in Equation (26) yields

$$T_{\mu\nu} = F_{\mu\rho}F^\rho{}_\nu + \frac{1}{4}g_{\mu\nu}F_{\rho\sigma}F^{\rho\sigma}, \tag{29}$$

from which we can obtain a covariant form of the classical electromagnetic Hamiltonian.

Note that, since in Equation (28) Σ is a spacelike hypersurface, n^μ must be timelike. Thus, there exists a frame in which $n_\mu J^\mu = n_0 J^0$, and as this is a scalar this is valid in any frame. We also have that $J^\mu = T_0^\mu$, so we seek T_0^0. On a static spacetime $T_0^0 = g^{00} T_{00}$, so

$$T_0^0 = \frac{1}{2}\left(\mathbf{E}^2 + \mathbf{B}^2\right), \qquad (30)$$

where in the intermediate step we have used the Minkowski field strength tensor, as the quantities are scalars. Hence we obtain the electromagnetic field Hamiltonian

$$H = \int_\Sigma d^3x \, \frac{1}{2}\left(\mathbf{E}^2 + \mathbf{B}^2\right) \sqrt{|\gamma|} n_0 K^0. \qquad (31)$$

This result is consistent with the literature [47], and reduces to the familiar expression in Equation (15) in Minkowski space.

For the covariant analogue of the quantum field Hamiltonian, we note that the field Hamiltonian used in the Minkowskian gauge-independent scheme, given in Equation (11), has a similar functional form to the Hamiltonian for a quantised scalar field; they are identical up to labelling and choice of integration measure. It has been established by Friis et al. [16] that the propagation of transverse electromagnetic field modes can be well approximated by such an uncharged field, and this technique has been used to determine the effects of spacetime curvature on satellite-based quantum communications and to make metrology predictions [23,26]. In the following, we use this approximation to justify the form of the electromagnetic field Hamiltonian from that of a real scalar field with the equation of motion $(\Box + m^2)\phi = 0$. The Hamiltonian density on a static manifold with Killing time t is

$$\mathcal{H} = \frac{\sqrt{|g|}}{2}\left(\partial^t\phi\partial_t\phi - \partial^i\phi\partial_i\phi + \frac{1}{2}m^2\phi^2 + \frac{1}{2}\xi R\phi\right). \qquad (32)$$

The final term pertains to the coupling between the spacetime background and the field. Given we just seek to study photons propagating on some curved background and are ignoring their back-reaction on the geometry, we can choose $\xi = 0$. This is known as the minimal coupling approximation.

Since on static spacetimes the d'Alembertian permits separable solutions, we can write $\phi = \psi_{\omega_\mathbf{k}}(\mathbf{x}) e^{\pm i E_{\omega_\mathbf{k}} t}$ [48,52]. Here $\psi_{\omega_\mathbf{k}}, E_{\omega_\mathbf{k}}$ are the eigenstates of the Klein–Gordon operator $(\Box + m^2)$. Upon quantisation, the field operator for a real scalar field can now be written as a linear superposition of these modes with ladder operators $b_{\omega_\mathbf{k}}, b_{\omega_\mathbf{k}}^\dagger$ defining the Fock space. However, we must also account for the nonuniqueness of particle states in curved spacetimes. One set of Fock space operators is often not able to cover an entire spacetime, so we will include a sum over distinct sets of operators, $b_{\omega_\mathbf{k}}^{(i)}, b_{\omega_\mathbf{k}}^{(i)\dagger}$. Following Fulling [52], we introduce a measure, $\mu(\omega_\mathbf{k})$, such that if the eigenstates form a complete basis for the Hilbert space of states, allowing a general function to be written as $F(\mathbf{x}) = \int d\mu(\omega_\mathbf{k})\left(f(\omega_\mathbf{k})\psi_{\omega_\mathbf{k}}(\mathbf{x})\right)$, the inner product on the Hilbert space becomes

$$\langle F_1, F_2 \rangle = \int d^3x \sqrt{|g|} g^{tt} F_1^* F_2 = \int d\mu(\omega_\mathbf{k}) f_1^* f_2. \qquad (33)$$

With this measure, the Hamiltonian field operator for a minimally-coupled scalar field on any static spacetime can be written as [52]

$$\hat{H} = \int d\mu(\omega_\mathbf{k}) \sum_i E_{\omega_\mathbf{k}}^{(i)} \left[\hat{b}_{\omega_\mathbf{k}}^{(i)\dagger} \hat{b}_{\omega_\mathbf{k}}^{(i)} + \frac{1}{2}\delta(0)\right]. \qquad (34)$$

Thus using the approximation of Friis et al. [16], we obtain the same functional form for the free electromagnetic quantised Hamiltonian on any static spacetime. To incorporate the direction of propagation, we can instead label modes in the above expressions by their wave-vector satisfying $|\mathbf{k}| = \omega_{\mathbf{k}}$. Then the integration measure $\mu(\omega_{\mathbf{k}})$ can be taken as $d\mu(\omega_{\mathbf{k}}) = d^3k$. This applies since

$$F(\mathbf{x}) = \int d^3k \, (f(\mathbf{k}) \psi_{\mathbf{k}}(\mathbf{x})) \tag{35}$$

and the inner product of two such functions is

$$\begin{aligned}
\langle F_{\mathbf{k}}, F_{\mathbf{k}'} \rangle &= \int d^3k \int d^3k' \int d^3x \, \sqrt{|g|} g^{tt} f_{\mathbf{k}}^* f_{\mathbf{k}'} \psi_{\mathbf{k}}^* \psi_{\mathbf{k}'} \\
&= \int d^3k \int d^3k' \, f_{\mathbf{k}}^* f_{\mathbf{k}'} \langle \psi_{\mathbf{k}}^*, \psi_{\mathbf{k}'} \rangle \\
&= \int d^3k \, f_{\mathbf{k}}^* f_{\mathbf{k}}.
\end{aligned} \tag{36}$$

To obtain the third line we have used that $\psi_{\mathbf{k}}$ and $\psi_{\mathbf{k}'}$ are eigenstates of a self-adjoint operator [52]. Physical photon modes will also be indexed by their transverse polarisation, so we also introduce an additional mode label for the polarisation λ. Thus, in all, for a minimally-coupled electromagnetic field on a static Lorentzian manifold the quantised field Hamiltonian for the Fock space defined in Equation (9) can be taken as

$$\hat{H} = \sum_{\lambda=1,2} \int d^3k \left(\sum_i \omega_{\mathbf{k}}^{(i)} \hat{b}_{\mathbf{k}\lambda}^{(i)\dagger} \hat{b}_{\mathbf{k}\lambda}^{(i)} + H_0 \right). \tag{37}$$

Other than the sum over distinct sectors, this result is no different from its Minkowskian counterpart; this has only been possible with careful considerations of the static curved background.

3.3.3. Electromagnetic Field Observables

The classical Hamiltonian remains quadratic in the electric and magnetic fields, and the quantised field Hamiltonian is still quadratic in the ladder operators. As is demonstrated above, this will continue to be the case for any static spacetime, as it was in the Minkowskian case of Section 2.3. In nonstatic spacetimes, the lack of a conserved local energy introduces ambiguity into our definition of the Hamiltonian and the scheme may no longer apply. Since the Hamiltonian is quadratic in both the field observables and the ladder operators, we can again make the ansatz that the electromagnetic field operators are linear superpositions of creation and annihilation operators. Assuming that the Hamiltonian and field operators retain the same relationships with one another as their classical counterparts guarantees the validity of this linear superposition, since there must exist a linear transformation between any two sets of variables if a quantity (the Hamiltonian) can be independently written as a quadratic function of each set. Our linear superposition of creation and annihilation operators takes as coefficients the negative and positive frequency modes respectively with respect to the future-directed timelike killing vector K^μ. The only modification we propose is the addition of a sum over spacetime sectors as introduced in the previous section. Including such flexibility will be essential in Section 4 when we quantise the electromagnetic field in an accelerated frame.

Thus the ansatz for the field operators becomes

$$\begin{aligned}
\hat{\mathbf{E}} &= \sum_{\lambda=1,2} \int d^3k \left(\sum_i p_{\mathbf{k}\lambda}^{(i)} \hat{b}_{\mathbf{k}\lambda}^{(i)} + \text{H.c.} \right) \hat{\mathbf{e}}_\lambda, \\
\hat{\mathbf{B}} &= \sum_{\lambda=1,2} \int d^3k \left(\sum_i q_{\mathbf{k}\lambda}^{(i)} \hat{b}_{\mathbf{k}\lambda}^{(i)} + \text{H.c.} \right) (\hat{\mathbf{k}} \times \hat{\mathbf{e}}_\lambda),
\end{aligned} \tag{38}$$

where $p_{\mathbf{k}\lambda}$ and $q_{\mathbf{k}\lambda}$ are unknown positive frequency mode functions of all the spacetime coordinates, and $\hat{\mathbf{e}}_\lambda$ is a unit polarisation vector orthogonal to the direction of the wave's propagation at a point x

in the spacetime. To determine the unknown mode functions, we demand that the expectation values of the operators satisfy the form of Maxwell's equations explicit in **E** and **B** that derives from

$$\frac{1}{\sqrt{|g|}}\partial_\mu\left(\sqrt{|g|}F^{\mu\nu}\right) = 0 \quad \text{and} \quad \partial_{[\sigma}F_{\mu\nu]} = 0. \tag{39}$$

In general, this could be highly nontrivial and is indeed the greatest obstacle to a simple implementation of the scheme. Solving wave equations on curved spacetimes is a difficult task [49], so we would like to again follow the Minkowskian scheme and simplify the task by using a Heisenberg equation of motion.

To get around the manifest noncovariance of Equation (13), we note that since $\hat{H}$ generates a unitary group that implements time translation symmetry on the Fock space, the equation is a geometric expression of the fact that time evolution of operators is generated by the system's Hamiltonian [52]. Considering the effect of an infinitesimal Poincaré transformation on an observable, $\hat{\mathcal{O}}$ thus gives

$$\partial_\mu \hat{\mathcal{O}} = -i[\hat{\mathcal{O}}, \hat{P}_\mu], \tag{40}$$

from which Equation (13) can be obtained as the 0th component [39,71,72]. Generalising this expression to curved spacetimes is then a simple matter of applying the minimal-coupling principle, giving

$$\nabla_\mu \hat{\mathcal{O}} = -i[\hat{\mathcal{O}}, \hat{P}_\mu]. \tag{41}$$

However, it is common to only consider evolution due to the Hamiltonian, in which case the Heisenberg equation is made covariant by using a proper time derivative to give [73–75]

$$\frac{d\hat{\mathcal{O}}}{d\tau} = -i[\hat{\mathcal{O}}, \hat{H}]. \tag{42}$$

Both approaches are used in the literature as covariant generalisations of the Heisenberg equation, yet they do not immediately appear to give the same results. To connect the two, we multiply Equation (41) by a tangent vector,

$$U^\mu \nabla_\mu \hat{\mathcal{O}} = -i\left[\hat{\mathcal{O}} U^\mu \hat{P}_\mu - U^\mu \hat{P}_\mu \hat{\mathcal{O}}\right], \tag{43}$$

where we have assumed that it commutes with all the operators. Along a curve x^α the directional derivative of any given tensor $\mathcal{T}$ is $\frac{d\mathcal{T}}{d\lambda} = \frac{dx^\alpha}{d\lambda}\nabla_\alpha \mathcal{T} = U^\alpha \nabla_\alpha \mathcal{T}$, where λ is any affine parameter. The case $\lambda = \tau$ promotes U^μ to the four velocity. For a particle on a stationary spacetime, in its rest frame $U_\mu P^\mu = H$, and as this is a scalar this holds in any frame. Thus one obtains Equation (42), which is the proper time covariant Heisenberg equation of motion.

Our generalised quantisation scheme will apply this covariant Heisenberg equation to the expectation value $\langle\hat{\mathcal{O}}\rangle$ of a general state in the photon Fock space $|\psi\rangle$,

$$\nabla_0 \langle\hat{\mathcal{O}}\rangle = -i\langle[\hat{\mathcal{O}}, \hat{H}]\rangle. \tag{44}$$

This gives the temporal evolution in the wave equations resulting from Equation (39), where $\hat{H}$ is taken as the field Hamiltonian of Equation (37). If the form of Maxwell's equations on the spacetime can be obtained and solved for the expectation values of the field operators using this procedure, the constant terms are determined by demanding that

$$\hat{H} \equiv \frac{1}{2}\int_\Sigma d^3x \left(\hat{\mathbf{E}}^2 + \hat{\mathbf{B}}^2\right)\sqrt{\gamma}n_0 K^0 \tag{45}$$

on the spacelike hypersurface Σ. As the integration over this hypersurface is independent of the choice of surface and is constant, this holds for all time. In this manner, the unknown modes in Equation (38) can be determined and the electromagnetic field on a static, 4-dimensional Lorentzian manifold can be quantised.

3.3.4. Summary of Scheme

Let us reflect on our construction. We have taken the Minkowskian gauge-independent electromagnetic field quantisation scheme in Section 2.3 and lifted it onto a static Lorentzian manifold with metric $g_{\mu\nu}$. Assuming the existence of detectable photons, the presence of a global timelike Killing vector allowed the definition of positive and negative frequency modes and thus the introduction of a well-defined particle Fock space, with general photon states labelled by their physical polarisation λ and wave-vector **k**. We introduced a ladder-operator structure for the Fock space, and using the approximation of Friis et al. [16] argued that this Fock space is associated with the field Hamiltonian of Equation (37) for minimal coupling to the background geometry.

The fact that both the field Hamiltonian $\hat{H}$ and the classical Hamiltonian H of Equation (31) were quadratic in the ladder operators or field strengths respectively allowed the proposal of a linear ansatz for the electric and magnetic field operators in terms of unknown wave modes. The scheme is then restricted to the specific manifold in question by demanding that the expectation values of these operators satisfy the modified Maxwell equations deriving from Equation (39), which introduces an explicit metric dependence to the scheme. To facilitate solving the potentially nontrivial Maxwell equations we use a form of the covariant Heisenberg equation, which we expect from work in Minkowski space to then uniquely determine the functional form of the modes in the operator ansatz. To determine all constants in these modes we demand that if we promote the classical Hamiltonian to an operator, upon substitution of the field operators the field Hamiltonian is regained.

By building off an already explicitly gauge-independent scheme, our method has the advantage of offering a gauge-independent and covariant route to the derivation of the Hamiltonian $\hat{H}$ and the electric and magnetic field observables, $\hat{\mathbf{E}}$ and $\hat{\mathbf{B}}$, respectively, on curved spacetimes. However, so far the only justification we have that this field quantisation scheme will give a physical result is based on its progenitor in Minkowski space. To test the consistency of our approach with other field quantisation schemes, we now consider a specific non-Minkowskian spacetime as an example and show that standard physical results are reproduced.

4. Electromagnetic Field Quantisation in an Accelerated Frame

In this section we apply the general formalism developed above to a specific example: 1-dimensional acceleration in Minkowski space. This situation is interesting as the noninertial nature of this motion leads to observers having different notions of particle states, and is thus often considered first in developments of quantum field theory in curved spacetime. It is also the situation most easily accessible to experimental tests. We must note that Soldati and Specchia [34] have emphasised photon propagation in accelerated frames remains conceptually nontrivial due to the separation of physical and nonphysical polarisation modes arising from standard quantisation techniques. Here we avoid these issues by only considering motion in the direction of acceleration (1D propagation) [33,34], and also by avoiding the use of canonical quantisation and immediately considering the physical degrees of freedom.

4.1. Rindler Space

An observer in Minkowski space $\mathbb{M}$ accelerating along a one-dimensional line with proper acceleration α appears to an inertial observer to travel along a hyperbolic worldline

$$x^\mu = \left(\frac{1}{\alpha}\sinh(\alpha\tau), \frac{1}{\alpha}\cosh(\alpha\tau), 0, 0\right), \quad x^\mu x_\mu = -\frac{1}{\alpha^2}, \tag{46}$$

where τ is the accelerating observer's proper time. As the proper acceleration $\alpha \to \infty$, the hyperbolic worldline of Equation (46) becomes asymptotic to the null lines of $\mathbb{M}$, $x = t$ for $t > 0$ and $x = -t$ for $t < 0$. The interior region in which the hyperbola resides is defined by $|t| < x$ and is called the Right Rindler wedge (RR); if $|t| < -x$ we have the Left Rindler wedge (LR). The union of both wedges yields the Rindler space $\mathcal{R}$, which is a static globally hyperbolic spacetime [58].

More concretely, we can obtain Rindler space by the coordinate transformation

$$t = \pm\rho \sinh(\alpha\zeta), \quad x = \pm\rho \cosh(\alpha\zeta), \quad y = y, \quad z = z, \tag{47}$$

where we call the coordinates (ζ, ρ, y, z) polar Rindler coordinates, with positive signs labelling points in RR and negative signs labelling those in LR [76]. In this coordinate system, the metric associated with the frame of accelerating observer O' is [34,56,62]

$$ds^2 = \alpha^2 \rho^2 d\zeta^2 - d\rho^2 - dy^2 - dz^2. \tag{48}$$

The right Rindler wedge is covered by the set of all uniformly accelerated motions such that $\alpha^{-1} \in \mathbb{R}^+$, and the boundaries of Rindler space are Cauchy horizons for the motion of O' [61,63].

Many studies of this spacetime choose to introduce conformal Rindler coordinates (ξ, η, y, z) [58,76], defined by the coordinate transformation

$$t = \pm a^{-1} e^{a\xi} \sinh(a\eta), \quad x = \pm a^{-1} e^{a\xi} \cosh(a\eta), \quad y = y, \quad z = z, \tag{49}$$

where $a \in \mathbb{R}$ is a positive constant such that $ae^{-a\xi} = \alpha$, so the proper time τ of O' relates to η as $\tau = e^{a\xi}\eta$. The two coordinate systems for $\mathcal{R}$ hence relate as

$$\rho = a^{-1} e^{a\xi} \quad \text{and} \quad \alpha\zeta = a\eta. \tag{50}$$

Lines of constant Rindler coordinates are shown in Figure 1. Rindler space can thus also be associated with the metric line element

$$ds^2 = e^{2a\xi}(d\eta^2 - d\xi^2) - dy^2 - dz^2. \tag{51}$$

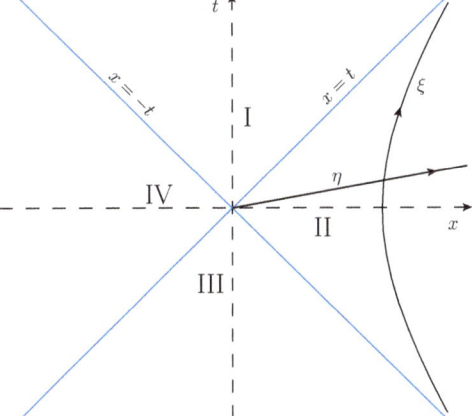

Figure 1. Depiction of a 2-dimensional Minkowski space $\mathbb{M}$. Regions I and III are the future and past light cones of the observer O at the origin, while regions II and IV are the right Rindler wedge (RR) and left Rindler wedge (LR) respectively. The worldline of a uniformly accelerated observer with acceleration α is the displayed line of constant conformal Rindler coordinate ξ.

These coordinates are useful because worldlines with $\zeta = 0$ have constant acceleration $a = \alpha$ [58].

From the discussion of Killing vectors in Section 3.3, it is immediate that since the metric components are independent of ζ or η in the respective coordinate systems, $\partial_\eta \equiv \frac{\alpha}{a}\partial_\zeta$ is a Killing field for $\mathcal{R}$, and moreover the field is timelike. However in LR the field is orientated in the past time direction, so the future-directed timelike killing vector in this wedge is $\partial_{(-\eta)} = -\partial_\eta \equiv -\frac{\alpha}{a}\partial_\zeta$. To deal with this when considering wave propagation, one must introduce two disjoint sets of positive frequency modes $f_k^{(i)}$, $i = L, R$. These satisfy

$$\partial_\eta f_k^{(R)} = -i\omega_k f_k^{(R)} \quad \text{and} \quad -\partial_\eta f_k^{(L)} = -i\omega_k f_k^{(L)}, \tag{52}$$

so each set is positive frequency with respect to its appropriate future-directed timelike Killing vector. These sets and their conjugates form a complete basis for solutions of the wave equation on $\mathcal{R}$ [48,51].

As a region of Minkowski space Rindler space is a flat spacetime with no matter content [64]. Despite this, because of the spacetime's noninertial nature covariant considerations must be applied when working in $\mathcal{R}$. For example the naïve divergence $\partial_\mu A^\mu \neq \partial^\mu A_\mu$ as required by Lorentz invariance, and we have non-zero Christoffel symbols

$$\Gamma^\zeta_{\zeta\zeta} = \Gamma^\zeta_{\eta\eta} = \Gamma^\eta_{\eta\zeta} = \Gamma^\eta_{\zeta\eta} = a. \tag{53}$$

With the Christoffel symbols covariant derivatives ∇_μ can be taken, and the timelike Killing vector fields ∂_η and $\partial_{(-\eta)}$ can be shown to formally satisfy Equation (20).

4.2. Electromagnetism in Rindler Space

To apply our covariant gauge-independent quantisation scheme to accelerating frames, we need to consider classical electromagnetism in Rindler space. Our starting point, the field strength tensor, takes the standard form

$$F^\mathcal{R}_{\mu\nu} = \begin{pmatrix} 0 & E^1_\mathcal{R} & E^2_\mathcal{R} & E^3_\mathcal{R} \\ -E^1_\mathcal{R} & 0 & -B^3_\mathcal{R} & B^2_\mathcal{R} \\ -E^2_\mathcal{R} & B^3_\mathcal{R} & 0 & -B^1_\mathcal{R} \\ -E^3_\mathcal{R} & -B^2_\mathcal{R} & B^1_\mathcal{R} & 0 \end{pmatrix}. \tag{54}$$

The explicit relations between the Rindler fields and those in Minkowski space are given in Appendix A. These relations are taken to define the fields in the accelerated frame. For the Maxwell equation we need the contravariant field strength tensor $F^{\mu\nu} = g^{\mu\sigma}g^{\nu\rho}F_{\sigma\rho}$. Because of the metric contractions this is explicitly coordinate dependent. In conformal coordinates we have

$$F^{\mu\nu}_\mathcal{R} = \begin{pmatrix} 0 & -E^1_\mathcal{R}e^{-4a\zeta} & -E^2_\mathcal{R}e^{-2a\zeta} & -E^3_\mathcal{R}e^{-2a\zeta} \\ E^1_\mathcal{R}e^{-4a\zeta} & 0 & -B^3_\mathcal{R}e^{-2a\zeta} & B^2_\mathcal{R}e^{-2a\zeta} \\ E^2_\mathcal{R}e^{-2a\zeta} & B^3_\mathcal{R}e^{-2a\zeta} & 0 & -B^1_\mathcal{R} \\ E^3_\mathcal{R}e^{-2a\zeta} & -B^2_\mathcal{R}e^{-2a\zeta} & B^1_\mathcal{R} & 0 \end{pmatrix}. \tag{55}$$

The polar coordinate form of this equation can be found in Appendix A.

The Maxwell equations that incorporate the spacetime's nontrivial geometry now follow from Equation (39). In Rindler space and conformal coordinates, $g = -e^{4a\zeta}$. Thus we obtain

$$\begin{aligned} e^{-2a\zeta}\partial_\zeta E^1_\mathcal{R} - 2a E^1_\mathcal{R} e^{-2a\zeta} + \partial_y E^2_\mathcal{R} + \partial_z E^3_\mathcal{R} &= 0, \\ e^{-2a\zeta}\partial_\eta E^1_\mathcal{R} &= \partial_y B^3_\mathcal{R} - \partial_z B^2_\mathcal{R}, \\ \partial_\eta E^2_\mathcal{R} &= e^{2a\zeta}\partial_z B^1_\mathcal{R} - \partial_\zeta B^3_\mathcal{R}, \\ \partial_\eta E^3_\mathcal{R} &= \partial_\zeta B^2_\mathcal{R} - e^{2a\zeta}\partial_y B^1_\mathcal{R}. \end{aligned} \tag{56}$$

The set of equations deriving from the Bianchi identity are exactly the same as in flat space; these are listed in Appendix A, along with the full Maxwell equations in polar coordinates.

4.3. Field Quantisation in Rindler Space

Knowing how classical electric and magnetic amplitudes evolve in Rindler space, we are now in a position to derive the Hamiltonian $\hat{H}$ and the electric and magnetic field observables, $\hat{\mathbf{E}}$ and $\hat{\mathbf{B}}$, respectively, of the quantised electromagnetic field in Rindler space $\mathcal{R}$. For simplicity, we are only interested in photons which propagate along one spatial dimension. Suppose they travel along the ζ axis in conformal or along the ρ axis in polar coordinates, which from Equation (50) are proportional and thus equivalent. Thus photon modes will have a wave-number k and a polarisation $\lambda = 1, 2$ as their labels. Working in only one dimension, we have avoided the necessity to introduce more complicated polarisations [34].

Unfortunately, the general states in Equation (9) are complicated in $\mathcal{R}$ by the existence of different future-directed timelike killing vectors in the two Rindler wedges, with ∂_η in RR and $-\partial_\eta$ in LR. Hence there need to be two sets of positive frequency modes for solutions of the wave equation on the spacetime. There will thus be two distinct Fock spaces representing the particle content in LR and RR. A general particle number state for light propagating in one dimension in $\mathcal{R}$ will hence be

$$\bigotimes_{\lambda=1,2} \bigotimes_{k=-\infty}^{\infty} \left| n^L_{k\lambda}, n^R_{k\lambda} \right\rangle, \qquad (57)$$

with $n^L_{k\lambda}$ being the number of photons in LR and $n^R_{k\lambda}$ being the number of photons in RR. Thus the physical energy eigenstates are in general degenerate and the Hamiltonian must satisfy

$$\hat{H} \left| n^L_{k\lambda}, n^R_{k\lambda} \right\rangle = \left[\omega_\mathbf{k}(n^L_{k\lambda} + n^R_{k\lambda}) + H_0 \right] \left| n^L_{k\lambda}, n^R_{k\lambda} \right\rangle, \qquad (58)$$

with integer values for both $n^L_{k\lambda}$ and $n^R_{k\lambda}$. This suggests that the field Hamiltonian $\hat{H}$ of Equation (37) has to be expressed in terms of independent ladder operators for both wedges. Hence, it can be written as

$$\hat{H} = \sum_{\lambda=1,2} \int_{-\infty}^{\infty} dk \left[\omega_\mathbf{k} (\hat{b}^{R\dagger}_{k\lambda} \hat{b}^R_{k\lambda} - \hat{b}^{L\dagger}_{k\lambda} \hat{b}^L_{k\lambda}) + H_0 \right], \qquad (59)$$

where the $E_{\omega_\mathbf{k}}$ factor of Equations (34) and (52) give the relative sign between the left and right sectors. As we are considering photons propagating along ζ or ρ, and photons are electromagnetic waves, the electric and magnetic fields must be in the transverse spatial dimensions y, z that are unaffected by the acceleration and thus identical to their Minkowski counterparts. As described in Section 2.3, the polarisation basis states correspond to choices of these fields. Here we choose

$$\mathbf{E}, \mathbf{B} = \begin{cases} (0, E, 0), & (0, 0, B) & \lambda = 1 \\ (0, 0, E), & (0, -B, 0) & \lambda = 2, \end{cases} \qquad (60)$$

where E and B are scalar functions of (ζ, ρ) or (η, ξ). With this choice of fields, the Rindler–Maxwell equations of Equation (56) reduce to

$$\partial_\eta E = -\partial_\xi B, \quad \partial_\xi E = -\partial_\eta B, \qquad (61)$$

for conformal Rindler coordinates, and from Equation (A7) to

$$\frac{1}{\rho^2 \alpha^2} \partial_\zeta E = -\left(\partial_\rho B + \frac{1}{\rho} B \right), \quad \partial_\rho E = -\partial_\zeta B, \qquad (62)$$

for polar coordinates. Both sets of equations hold in both LR and RR. The conformal expressions are now identical to the 1D Minkowski propagation considered in [35]. It should be emphasised that the apparent simplicity is a result of demanding 1-dimensional propagation along the accelerated spatial axis and choosing convenient polarisations.

The noninertial nature of Rindler space still requires care; recall from Equation (31) that to determine the classical electromagnetic Hamiltonian, we require a timelike Killing vector field. We must also choose a spacelike hypersurface Σ with normal vector n^μ and induced metric γ_{ij} to integrate over. In conformal Rindler coordinates, we know that the timelike Killing vector field is $K = \partial_\eta$, so $K^\mu = \delta^\mu_\eta$. Choosing Σ as being the hypersurface defined by $\eta = 0$ allows us to continue using the spatial coordinates $x^i = (\xi, y, z)$. Hence, the full conformal Rindler metric of Equation (51) implies $\gamma = \det(\gamma_{ij}) = e^{-2a\xi}$. Finally, since Σ is spacelike, n^μ is normalised to $+1$, so

$$1 = g_{\mu\nu} n^\mu n^\nu = e^{2a\xi} \left(n^0\right)^2, \tag{63}$$

giving $n^0 = e^{-a\xi}$ [48]. Hence the Hamiltonian in Rindler space is

$$\begin{aligned} H &= \frac{1}{2} \int d\xi \left(\mathbf{E}^2 + \mathbf{B}^2\right) e^{a\xi} e^{-a\xi} \delta^\eta_\eta \\ &= \frac{1}{2} \int d\xi \left(\mathbf{E}^2 + \mathbf{B}^2\right), \end{aligned} \tag{64}$$

so the initial apparent simplicity holds.

Following our general prescription, we again make the ansatz that the field operators are linear superpositions of the relevant ladder operators. As we are considering 1-dimensional propagation with the electric and magnetic field vectors $\mathbf{E}$ and $\mathbf{B}$, respectively, as specified in Equation (60), we need only apply the ansatz to the scalar components E and B for quantisation, giving

$$\begin{aligned} \hat{E} &= \sum_{\lambda=1,2} \int_{-\infty}^{\infty} dk \left(p^L_{k\lambda} \hat{b}^L_{k\lambda} + p^R_{k\lambda} \hat{b}^R_{k\lambda} + \text{H.c.} \right), \\ \hat{B} &= \sum_{\lambda=1,2} \int_{-\infty}^{\infty} dk \left(q^L_{k\lambda} \hat{b}^L_{k\lambda} + q^R_{k\lambda} \hat{b}^R_{k\lambda} + \text{H.c.} \right), \end{aligned} \tag{65}$$

where $p^i_{k\lambda}$ and $q^i_{k\lambda}$ are unknown functions of (η, ξ), and $i = L, R$ for LR and RR respectively. Since the left and the right wedges of $\mathcal{R}$ are causally disjoint, we can demand that modes in different wedges are orthogonal with respect to the inner product in Equation (36) [48]. Explicitly this yields

$$\begin{aligned} \langle p^L_{k\lambda}, p^R_{k'\lambda'} \rangle &= \int_{-\infty}^{\infty} dk \, p^{*L}_{k\lambda} p^R_{k'\lambda'} = 0, \\ \langle p^{*L}_{k\lambda}, p^R_{k'\lambda'} \rangle &= \int_{-\infty}^{\infty} dk \, p^L_{k\lambda} p^R_{k'\lambda'} = 0 \end{aligned} \tag{66}$$

with similar expressions for $q^i_{k\lambda}$. To determine all the modes, we follow the recipe of Section 3.3 and demand that the expectation values of these field operators satisfy Equations (61) and (62).

From now on we will work in the conformal Rindler coordinates (η, ξ) due to the wonderful simplicity of their Maxwell equations. One could of course also use polar coordinates, and indeed one can show that this yields the same results in this set for the case $a = \alpha$. To determine temporal evolution we use the Heisenberg equation, which, as the time coordinate is η in this system and our observables $\hat{O}$ are scalars, is

$$\partial_\eta \hat{O} = -i[\hat{O}, \hat{H}]. \tag{67}$$

Following our prescription, we compare expectation values of the ladder operators for spatial derivatives and time evolution from Heisenberg's equation by using our form of Maxwell's equations. In this case, using Equations (61) this procedure gives the relations

$$\partial_\xi q^i_{k\lambda} = i\omega_\mathbf{k} p^i_{k\lambda}, \tag{68}$$

$$\partial_\xi p^i_{k\lambda} = i\omega_\mathbf{k} q^i_{k\lambda}. \tag{69}$$

Solving for $p^i_{k\lambda}$ we of course just obtain the wave equation, $\left(\partial_\xi^2 + k^2\right) p^i_{k\lambda} = 0$, when we consider free, on-shell photons with $k^2 = \omega_\mathbf{k}^2$. This equation admits separable solutions $p^i_{k\lambda} = \chi^i_{k\lambda}(\eta) P^i_{k\lambda}(\xi)$, so as there are no temporal derivatives we lose all temporal information. Writing the spatial solution is trivial:

$$P^i_{k\lambda} = J^i_\lambda e^{ik\xi} + K^i_\lambda e^{-ik\xi}, \tag{70}$$

where $J^i_\lambda, K^i_\lambda \in \mathbb{C}$. To determine the temporal dependence of $\chi_{k\lambda}(\eta)$ we use that positive frequency Rindler modes must satisfy Equation (52). The two modes $p^L_{k\lambda}$ and $p^R_{k\lambda}$ must both be positive frequency with respect to the future-direction of ∂_η as they are coefficients of annihilation operators [40]. Thus the difference between them will be in their time dependence. This gives that we must have

$$\chi^L_{k\lambda} = e^{i\omega_\mathbf{k}\eta}, \quad \chi^R_{k\lambda} = e^{-i\omega_\mathbf{k}\eta}. \tag{71}$$

This difference is a direct result of the two Rindler wedges having different future-directed timelike Killing vectors. Thus, in all, we have

$$\begin{aligned} p^R_{k\lambda}(\eta,\xi) &= U^R_\lambda e^{i(k\xi-\omega_\mathbf{k}\eta)} + V^R_\lambda e^{-i(k\xi+\omega_\mathbf{k}\xi)}, \\ p^L_{k\lambda}(\eta,\xi) &= U^L_\lambda e^{i(k\xi+\omega_\mathbf{k}\eta)} + V^L_\lambda e^{-i(k\xi-\omega_\mathbf{k}\eta)}. \end{aligned} \tag{72}$$

We can then easily obtain the $q^i_{k\lambda}$ solutions from Equation (68) as

$$\begin{aligned} q^R_{k\lambda}(\eta,\xi) &= \frac{k}{\omega_\mathbf{k}} \left[U^R_\lambda e^{i(k\xi-\omega_\mathbf{k}\eta)} - V^R_\lambda e^{-i(k\xi+\omega_\mathbf{k}\eta)} \right], \\ q^L_{k\lambda}(\eta,\xi) &= \frac{k}{\omega_\mathbf{k}} \left[U^L_\lambda e^{i(k\xi+\omega_\mathbf{k}\eta)} - V^L_\lambda e^{-i(k\xi-\omega_\mathbf{k}\eta)} \right]. \end{aligned} \tag{73}$$

We now seek to determine the unknown coefficients in these expressions. Similarly to Section 2.3, first note that wave modes propagating in the positive ξ direction in $\mathcal{R}$ should be functions of $(k\xi - \omega_\mathbf{k}\eta)$ in RR where ∂_η is the future-directed timelike Killing vector, and functions of $(k\xi + \omega_\mathbf{k}\eta)$ in LR where it is $-\partial_\eta$. Similarly, modes propagating in the negative ξ direction should be functions of $(k\xi + \omega_\mathbf{k}\eta)$ in RR and functions of $(k\xi - \omega_\mathbf{k}\eta)$ in LR. These conditions imply $V^R = V^L = 0$.

We then determine the remaining constants by demanding that the classical and the quantised field Hamiltonians are equivalent, as in Equation (45). Since $\hat{H}$ is quadratic in the electric and magnetic field operators, we obtain cross terms between LR and RR modes during the calculation. Integrating over such terms gives the inner products in Equation (66), but as modes in the different wedges are orthogonal these terms are identically 0, so there are no physical cross terms. Then after some algebra and relying on the integral definition of the delta function, we arrive at

$$\hat{H} = 2\pi \sum_{\lambda=1,2} \int_{-\infty}^{\infty} dk \left[|U^R_\lambda|^2 \left(2\hat{b}^{\dagger R}_{k\lambda} \hat{b}^R_{k\lambda} + \delta(0) \right) + |U^L_\lambda|^2 \left(2\hat{b}^{\dagger L}_{k\lambda} \hat{b}^L_{k\lambda} + \delta(0) \right) \right], \tag{74}$$

where we have used the commutation relations in Equation (25). As in Section 2.3, to finally determine the constant terms and zero-point energy we compare with Equation (59) which yields

$$|U_\lambda^R|^2 = \frac{\omega_\mathbf{k}}{4\pi}, \quad |U_\lambda^L|^2 = \frac{\omega_\mathbf{k}}{4\pi}, \quad H_0 = \int_{-\infty}^{\infty} dk \, \omega_\mathbf{k} \, \delta(0). \quad (75)$$

To obtain our final expressions for the electric and magnetic field operators we arbitrarily choose the phases of both U_λ^R and U_λ^L to give consistency with standard Minkowskian results, and multiply the electric field operator by polarisation unit vector $\hat{\mathbf{e}}_\lambda$ and the magnetic field operator by $\hat{\mathbf{k}} \times \hat{\mathbf{e}}_\lambda$. Thus, in all, we obtain the final results

$$\hat{\mathbf{E}} = i \sum_{\lambda=1,2} \int_{-\infty}^{\infty} dk \sqrt{\frac{\omega_\mathbf{k}}{4\pi}} \left[e^{i(k\xi - \omega_\mathbf{k}\eta)} \hat{b}_{k\lambda}^R + e^{i(k\xi + \omega_\mathbf{k}\eta)} \hat{b}_{k\lambda}^L + \text{H.c.} \right] \hat{\mathbf{e}}_\lambda,$$

$$\hat{\mathbf{B}} = -i \sum_{\lambda=1,2} \int_{-\infty}^{\infty} dk \sqrt{\frac{\omega_\mathbf{k}}{4\pi}} \left[e^{i(k\xi - \omega_\mathbf{k}\eta)} \hat{b}_{k\lambda}^R + e^{i(k\xi + \omega_\mathbf{k}\eta)} \hat{b}_{k\lambda}^L + \text{H.c.} \right] (\hat{\mathbf{k}} \times \hat{\mathbf{e}}_\lambda),$$

$$\hat{H} = \sum_{\lambda=1,2} \int_{-\infty}^{\infty} dk \, \omega_\mathbf{k} \left[\hat{b}_{k\lambda}^{\dagger R} \hat{b}_{k\lambda}^R + \hat{b}_{k\lambda}^{\dagger L} \hat{b}_{k\lambda}^L + \delta(0) \right]. \quad (76)$$

These three operators are very similar to the electric and magnetic field operators $\hat{\mathbf{E}}$ and $\hat{\mathbf{B}}$, respectively, and $\hat{H}$ in Equations (11) and (14) in Minkowski space. When moving in only one dimension, the orientation of the electric and magnetic field amplitudes is still pairwise orthogonal and orthogonal to the direction of propagation. However, the electromagnetic field has become degenerate and additional degrees of freedom which correspond to different Rindler wedges have to be taken into account in addition to the wave numbers k and the polarisations λ of the photons. Finally, instead of depending on kx, the electric and magnetic field observables now depend on $k\xi \pm \omega_\mathbf{k}\eta$, i.e., they depend not only on the position but also on the amount of time the observer has been accelerating in space and on their acceleration. Most importantly, Equation (76) can now be used as the starting point for further investigations into the quantum optics of an accelerating observer [5,36,46], and is expected to find immediate applications in relativistic quantum information [13–21,69].

4.4. The Unruh Effect

As an example and to obtain a consistency check, we now verify that our results give the well-established Unruh effect [55,56,58,59]. This effect predicts that an observer with uniform acceleration α in Minkowski space measures the Minkowski vacuum as being a pure thermal state with temperature

$$T_\text{Unruh} = \frac{\alpha}{2\pi}. \quad (77)$$

Deriving this result relies on being able to switch between modes in Minkowski and modes in Rindler space, which requires a Bogolubov transformation. This transformation allows us to switch between the modes of different coordinate frames and generally transforms a vacuum state to a thermal state [57,77]. For a field expansion in two complete sets of basis modes, $\phi = \sum_i \hat{a}_i f_i + \hat{a}_i^\dagger f_i^* = \sum_j \hat{b}_j g_j + \hat{b}_j^\dagger g_j$, this relates the modes as

$$g_i = \sum_j \alpha_{ij} f_j + \beta_{ij} f_j^*,$$

$$f_i = \sum_j \alpha_{ji}^* g_j - \beta_{ji} g_j^*, \quad (78)$$

where α_{ij} and β_{ij} are the Bogolubov coefficients [58]. Knowing these coefficients also allows the associated particle Fock spaces to be related,

$$\hat{a}_i = \sum_j \alpha_{ji} \hat{b}_j + \beta_{ji}^* \hat{b}_j^\dagger,$$
$$\hat{b}_i = \sum_j \alpha_{ij}^* \hat{a}_j - \beta_{ij}^* \hat{a}_j^\dagger. \tag{79}$$

For transforming between the Rindler and Minkowski Fock spaces, the coefficients can be calculated using coordinate relations in a method first introduced by Unruh [55].

Here our field modes are the expansions of the electric field operators in $\mathcal{R}$ and $\mathbb{M}$ with the Minkowski results taking the same functional form. Following the standard approach [48,51], our expressions for the field operators yield

$$\alpha_{LL} = \alpha_{RR} = \frac{1}{\omega_{\mathbf{k}}} \sqrt{\frac{1}{2\sinh(\frac{\pi\omega_{\mathbf{k}}}{a})}} \, e^{\frac{\pi\omega_{\mathbf{k}}}{2a}}$$
$$\beta_{LR} = \beta_{RL} = \frac{1}{\omega_{\mathbf{k}}} \sqrt{\frac{1}{2\sinh(\frac{\pi\omega_{\mathbf{k}}}{a})}} \, e^{-\frac{\pi\omega_{\mathbf{k}}}{2a}}. \tag{80}$$

These immediately give the following relationship between the ladder operators.

$$\hat{b}_{k\lambda}^R = \frac{1}{\omega_{\mathbf{k}}} \sqrt{\frac{1}{2\sinh(\frac{\pi\omega_{\mathbf{k}}}{a})}} \left(e^{\frac{\pi\omega_{\mathbf{k}}}{2a}} \hat{c}_{k\lambda}^R + e^{-\frac{\pi\omega_{\mathbf{k}}}{2a}} \hat{c}_{-k\lambda}^{L\dagger} \right),$$
$$\hat{b}_{k\lambda}^L = \frac{1}{\omega_{\mathbf{k}}} \sqrt{\frac{1}{2\sinh(\frac{\pi\omega_{\mathbf{k}}}{a})}} \left(e^{\frac{\pi\omega_{\mathbf{k}}}{2a}} \hat{c}_{k\lambda}^L + e^{-\frac{\pi\omega_{\mathbf{k}}}{2a}} \hat{c}_{-k\lambda}^{R\dagger} \right). \tag{81}$$

The $\hat{c}_{k\lambda}^i$ operators are associated with modes that can be purely expressed in terms of positive frequency Minkowski modes (from the form of the field operators in Cartesian coordinates). They must thus share the Minkowski vacuum, so $\hat{c}_k^R |0_\mathbb{M}\rangle = \hat{c}_k^L |0_\mathbb{M}\rangle = 0$. Because we possess the Bogolubov transformation between Minkowski and Rindler space, we can now evaluate particle states seen by an observer in $\mathcal{R}$, given by $\hat{b}_k^i$, in terms of a Minkowski Fock space given by c_k^i. In particular, evaluating the RR number operator on the Minkowski vacuum gives

$$\langle 0_\mathbb{M} | \hat{b}_k^{R\dagger} \hat{b}_k^R | 0_\mathbb{M} \rangle = \frac{1}{\omega_{\mathbf{k}}^2} \frac{\delta(0)}{\exp(\frac{2\pi\omega_{\mathbf{k}}}{a}) - 1}. \tag{82}$$

This energy expectation value is the same as the energy expectation value of a thermal Planckian state with temperature $\frac{a}{2\pi}$. For the case $a = \alpha$ this is the prediction that exactly constitutes the Unruh effect, and thus verifies that the results of our quantisation scheme match known theoretical predictions. Having $a \neq \alpha$ just corresponds to a redshift [48]. The external factor $1/\omega_{\mathbf{k}}^2$ is different to that for a standard scalar field; this is just a remnant of the different normalisation of our electric field operator and does not affect the physical prediction, with such factors indeed sometimes appearing in the literature [49].

5. Conclusions

This paper generalises the physically-motivated quantisation scheme of the electromagnetic field in Minkowski space [35] to static spacetimes of otherwise arbitrary geometry. As shown in Section 3, such a generalisation requires only minimal modification of the original quantisation scheme in flat space. In order to assess the validity of the presented generalised approach, we apply our findings in Section 4 to the well understood case of Rindler space: the relevant geometry for a uniformly accelerating observer. Since this reproduces the anticipated Unruh effect, it supports the hypothesis that our approach is a consistent approach to the quantisation of the electromagnetic field on curved spacetimes.

The main strength of our quantisation scheme is its gauge-independence, i.e., its nonreliance on the gauge-dependent potentials of more traditional approaches. Instead it relies only on the experimentally verified existence of electromagnetic field quanta. As such, our scheme provides a more intuitive approach to field quantisation, while still relying on well established concepts and constructions in quantum field theory in curved space. Given this and the applicability of our results to accelerating frames in an otherwise flat spacetime, it seems likely that our approach can also be used to model more complex, but experimentally accessible, situations with applications, for example, in relativistic quantum information.

The specific case of Rindler space, as considered in this paper, led to equations with straightforward analytic solutions. This will likely not be true in more general settings, where the necessary wave equations will be nontrivial and will possibly require approximation or numerical solution. This fact is partially mitigated by our use of the Heisenberg equation, thereby reducing the necessary calculation to an ordinary differential equation and commutation relation, rather than a partial differential equation. Furthermore, recall that the scheme laid out in this paper is a generalisation of that in flat space to the case of static curved spacetimes. This simplified the definition and construction of the quantisation scheme, due to our reliance on spacelike hypersurfaces. When applied to the more general case of stationary spacetimes, the correct prescription of the scheme becomes less clear and will require further theoretical development.

Author Contributions: B.M. wrote the original draft. A.B. initiated the project. A.B. and R.P. both provided guidance and supervision. D.H. and R.P. checked all of the calculations and helped to improve the manuscript. All authors contributed to the reviewing and editing of the paper.

Funding: B.M. acknowledges financial support from the Science and Technology Facilities Council STFC grant number ST/R504737/1. Moreover, A.B. acknowledges financial support from the Oxford Quantum Technology Hub for Networked Quantum Information Technology NQIT (grant number EP/M013243/1).

Acknowledgments: We would like to thank Lewis A. Clark, Jiannis K. Pachos and Jan Sperling for helpful discussions. Statement of compliance with EPSRC policy framework on research data: This publication is theoretical work that does not require supporting research data.

Conflicts of Interest: The authors declare no conflicts of interest. The funder had no role in the design of the study, the writing of the manuscript, or the decision to publish the results.

Appendix A. Further Results of Electromagnetism in Rindler Space

To define the electric and magnetic fields in Rindler space we apply coordinate transformations to the Minkowski field strength tensor,

$$F^{\mathcal{R}}_{\mu\nu} = \frac{\partial x^{\alpha}_{\mathrm{M}}}{\partial x^{\mu}_{\mathcal{R}}} \frac{\partial x^{\beta}_{\mathrm{M}}}{\partial x^{\nu}_{\mathcal{R}}} F_{\alpha\beta}, \tag{A1}$$

where $x^{\mu}_{\mathcal{R}}$ are the coordinates in Rindler space and x^{μ}_{M} are the coordinates used by an intertial observer. The Rindler electric and magnetic fields are defined as the elements of $F^{\mathcal{R}}_{\mu\nu}$. In polar and conformal coordinates this transformation is given by Equations (47) and (49), respectively, which readily give the Jacobian of the transformation as

$$J_{\mu}{}^{\alpha} \equiv \frac{\partial x^{\alpha}_{\mathrm{M}}}{\partial x^{\mu}_{\mathcal{R}}} = \begin{pmatrix} \pm\alpha\rho\cosh(\alpha\zeta) & \pm\sinh(\alpha\zeta) & 0 & 0 \\ \pm\alpha\rho\sinh(\alpha\zeta) & \pm\cosh(\alpha\zeta) & 0 & 0 \\ 0 & 0 & 1 & 0 \\ 0 & 0 & 0 & 1 \end{pmatrix} \tag{A2}$$

in polar Rinder coordinates and

$$J_\mu{}^\alpha \equiv \frac{\partial x_M^\alpha}{\partial x_\mathcal{R}^\mu} = \begin{pmatrix} \pm e^{a\zeta}\cosh(a\eta) & \pm e^{a\zeta}\sinh(a\eta) & 0 & 0 \\ \pm e^{a\zeta}\sinh(a\eta) & \pm e^{a\zeta}\cosh(a\eta) & 0 & 0 \\ 0 & 0 & 1 & 0 \\ 0 & 0 & 0 & 1 \end{pmatrix} \quad (A3)$$

in conformal Rindler coordinates, where upper signs refer to RR and lower signs to LR. Transforming the Minkowski field strength tensor in Equation (A1), we obtain $F_{\mu\nu}^\mathcal{R}$ in Equation (54), where the Rindler space elements are defined in either wedge by the transformations

$$\begin{aligned}
E_\mathcal{R}^1 &= E_M^1 \alpha\rho, \\
E_\mathcal{R}^2 &= \left(E_M^2 \alpha\rho \cosh(\alpha\zeta) - B_M^3 \sinh(\alpha\zeta)\right), \\
E_\mathcal{R}^3 &= \left(E_M^3 \alpha\rho \cosh(\alpha\zeta) + B_M^2 \sinh(\alpha\zeta)\right), \\
B_\mathcal{R}^1 &= B_M^1, \\
B_\mathcal{R}^2 &= \left(B_M^2 \cosh(\alpha\zeta) + E_M^3 \alpha\rho \sinh(\alpha\zeta)\right), \\
B_\mathcal{R}^3 &= \left(B_M^3 \cosh(\alpha\zeta) - E_M^2 \alpha\rho \sinh(\alpha\zeta)\right),
\end{aligned} \quad (A4)$$

in polar Rindler coordinates and

$$\begin{aligned}
E_\mathcal{R}^1 &= E_M^1 e^{2a\zeta}, \\
E_\mathcal{R}^2 &= \left(E_M^2 \cosh(a\eta) - B_M^3 \sinh(a\eta)\right) e^{a\zeta}, \\
E_\mathcal{R}^3 &= \left(E_M^3 \cosh(a\eta) + B_M^2 \sinh(a\eta)\right) e^{a\zeta}, \\
B_\mathcal{R}^1 &= B_M^1, \\
B_\mathcal{R}^2 &= \left(B_M^2 \cosh(a\eta) + E_M^3 \sinh(a\eta)\right) e^{a\zeta}, \\
B_\mathcal{R}^3 &= \left(B_M^3 \cosh(a\eta) - E_M^2 \sinh(a\eta)\right) e^{a\zeta},
\end{aligned} \quad (A5)$$

in conformal Rindler coordinates. While the conformal coordinate form of the field strength tensor is listed in Equation (55), that for polar coordinates, which equals

$$F_\mathcal{R}^{\mu\nu} = \begin{pmatrix} 0 & \frac{-E_\mathcal{R}^1}{\rho^2 \alpha^2} & \frac{-E_\mathcal{R}^2}{\rho^2 \alpha^2} & \frac{-E_\mathcal{R}^3}{\rho^2 \alpha^2} \\ \frac{E_\mathcal{R}^1}{\rho^2 \alpha^2} & 0 & -B_\mathcal{R}^3 & B_\mathcal{R}^2 \\ \frac{E_\mathcal{R}^1}{\rho^2 \alpha^2} & B_\mathcal{R}^3 & 0 & -B_\mathcal{R}^1 \\ \frac{E_\mathcal{R}^1}{\rho^2 \alpha^2} & -B_\mathcal{R}^2 & B_\mathcal{R}^1 & 0 \end{pmatrix}, \quad (A6)$$

was omitted. Then, since in polar coordinates, $g = -\rho^2 \alpha^2$, Equation (5) gives that the modified Maxwell equations in these coordinates are

$$\partial_\rho E^1_\mathcal{R} - \frac{1}{\rho} E^1_\mathcal{R} + \partial_y E^2_\mathcal{R} + \partial_z E^3_\mathcal{R} = 0,$$
$$\frac{1}{\rho^2 \alpha^2} \partial_\zeta E^1_\mathcal{R} = \partial_y B^3_\mathcal{R} - \partial_z B^2_\mathcal{R},$$
$$\frac{1}{\rho^2 \alpha^2} \partial_\zeta E^2_\mathcal{R} = \partial_z B^1_\mathcal{R} - \partial_\rho B^3_\mathcal{R} - \frac{1}{\rho} B^3_\mathcal{R},$$
$$\frac{1}{\rho^2 \alpha^2} \partial_\zeta E^3_\mathcal{R} = \partial_\rho B^2_\mathcal{R} + \frac{1}{\rho} B^2_\mathcal{R} - \partial_y B^1_\mathcal{R},$$
(A7)

while the Bianchi identity leads to

$$\partial_i B^i_\mathcal{R} = 0,$$
$$\partial_\zeta B^1_\mathcal{R} = \partial_z E^2_\mathcal{R} - \partial_y E^3_\mathcal{R},$$
$$\partial_\zeta B^2_\mathcal{R} = \partial_\rho E^3_\mathcal{R} - \partial_z E^1_\mathcal{R},$$
$$\partial_\zeta B^3_\mathcal{R} = \partial_y E^1_\mathcal{R} - \partial_\rho E^2_\mathcal{R},$$
(A8)

as in flat space. These equations also hold for conformal coordinates; one just replaces ζ with η and ρ with $\bar{\zeta}$.

References

1. Roychoudhuri, C.; Kracklauer, A.F.; Creath, K. *The Nature of Light: What Is a Photon?* CRC Press: Boca Raton, FL, USA, 2008.
2. Stokes, A. On Gauge Freedom and Subsystems in Quantum Electrodynamics. Ph.D. Thesis, School of Physics and Astronomy, University of Leeds, Leeds, UK, January 2014.
3. Andrews, D.L. Physicality of the photon. *J. Phys. Chem. Lett.* **2013**, *4*, 3878–3884. [CrossRef]
4. Barlow, T.M.; Bennett, R.; Beige, A. A master equation for a two-sided optical cavity. *J. Mod. Opt.* **2015**, *62*, S11–S20. [CrossRef] [PubMed]
5. Furtak-Wells, N.; Clark, L.A.; Purdy, R.; Beige, A. Quantising the electromagnetic field near two-sided semitransparent mirrors. *Phys. Rev.* **2018**, *97*, 043827. [CrossRef]
6. Nisbet-Jones, P.B.R.; Dilley, J.; Ljunggren, D.; Kuhn, A. Highly efficient source for indistinguishable single photons of controlled shape. *New J. Phys.* **2011**, *13*, 103036. [CrossRef]
7. Milburn, G.J.; Basiri-Esfahani, S. Quantum optics with one or two photons. *Proc. R. Soc. A* **2015**, *471*, 20150208. [CrossRef] [PubMed]
8. Kuhn, A.; Hennrich, M.; Rempe, G. Deterministic Single-Photon Source for Distributed Quantum Networking. *Phys. Rev. Lett.* **2002**, *89*, 067901. [CrossRef] [PubMed]
9. Pan, J.W.; Bouwmeester, D.; Daniell, M.; Weinfurter, H.; Zeilinger, A. Experimental test of quantum nonlocality in three-photon Greenberger-Horne-Zeilinger entanglement. *Nature* **2000**, *403*, 515–519. [CrossRef]
10. Eisaman, M.D.; Fan, J.; Migdall, A.; Polyakov, S.V. Invited review article: Single-photon sources and detectors. *Rev. Sci. Instrum.* **2011**, *82*, 071101. [CrossRef]
11. Giustina, M.; Versteegh, M.A.M.; Wengerowsky, S.; Handsteiner, J.; Hochrainer, A.; Phelan, K.; Steinlechner, F.; Kofler, J.; Larsson, J.A.; Abellan, C.; et al. Significant-Loophole-Free Test of Bell's Theorem with Entangled Photons. *Phys. Rev. Lett.* **2015**, *115*, 250401. [CrossRef]
12. Hensen, B.; Bernien, H.; Dréau, A.E.; Reiserer, A.; Kalb, N.; Blok, M.S.; Ruitenberg, J.; Vermeulen, R.F.L.; Schouten, R.N.; Abellan, C.; et al. Loophole-free Bell inequality violation using electron spins separated by 1.3 kilometres. *Nature* **2015**, *526*, 682–686. [CrossRef]
13. Villoresi, P.; Jennewein, T.; Tamburini, F.; Aspelmeyer, M.; Bonato, C.; Ursin, R.; Pernechele, C.; Luceri, V.; Bianco, G.; Zeilinger, A.; et al. Experimental verification of the feasibility of a quantum channel between space and earth. *New J. Phys.* **2008**, *10*, 033038. [CrossRef]

14. Vallone, G.; Bacco, D.; Dequal, D.; Gaiarin, S.; Luceri, V.; Bianco, G.; Villoresi, P. Experimental Satellite Quantum Communications. *Phys. Rev. Lett.* **2015**, *115*, 040502. [CrossRef] [PubMed]
15. Rideout, D.; Jennewein, T.; Amelino-Camelia, G.; Demarie, T.F.; Higgins, B.L.; Kempf, A.; Kent, A.; Laflamme, R.; Ma, X.; Mann, R.B.; et al. Fundamental quantum optics experiments conceivable with satellites reaching relativistic distances and velocities. *Class. Quantum Gravity* **2012**, *29*, 224011. [CrossRef]
16. Friis, N.; Lee, A.R.; Louko, J. Scalar, spinor, and photon fields under relativistic cavity motion. *Phys. Rev. D* **2013**, *88*, 064028. [CrossRef]
17. Friis, N.; Lee, A.R.; Truong, K.; Sabín, C.; Solano, E.; Johansson, G.; Fuentes, I. Relativistic quantum teleportation with superconducting circuits. *Phys. Rev. Lett.* **2013**, *110*, 113602. [CrossRef] [PubMed]
18. Ahmadi, M.; Bruschi, D.E.; Fuentes, I. Quantum metrology for relativistic quantum fields. *Phys. Rev. D* **2014**, *89*, 065028. [CrossRef]
19. Bruschi, D.E.; Sabín, C.; Kok, P.; Johansson, G.; Delsing, P.; Fuentes, I. Towards universal quantum computation through relativistic motion. *Sci. Rep.* **2016**, *6*, 18349. [CrossRef] [PubMed]
20. Kravtsov, K.S.; Radchenko, I.V.; Kulik, S.P.; Molotkov, S.N. Relativistic quantum key distribution system with one-way quantum communication. *Sci. Rep.* **2018**, *8*, 6102. [CrossRef]
21. Lopp, R.; Martín-Martínez, E. Light, matter, and quantum randomness generation: A relativistic quantum information perspective. *Opt. Commun.* **2018**, *423*, 29–47. [CrossRef]
22. Bruschi, D.E.; Fuentes, I.; Louko, J. Voyage to alpha centauri: Entanglement degradation of cavity modes due to motion. *Phys. Rev. D* **2012**, *85*, 061701. [CrossRef]
23. Bruschi, D.E.; Ralph, T.C.; Fuentes, I.; Jennewein, T.; Razavi, M. Spacetime effects on satellite-based quantum communications. *Phys. Rev. D* **2014**, *90*, 045041. [CrossRef]
24. Bruschi, D.E.; Sabín, C.; White, A.; Baccetti, V.; Oi, D.K.L.; Fuentes, I. Testing the effects of gravity and motion on quantum entanglement in space-based experiments. *New J. Phys.* **2014**, *16*, 053041. [CrossRef]
25. Calmet, X.; Dunningham, J. Transformation properties and entanglement of relativistic qubits under space-time and gauge transformations. *Phys. Rev. A* **2017**, *95*, 042309. [CrossRef]
26. Bruschi, D.E.; Datta, A.; Ursin, R.; Ralph, T.C.; Fuentes, I. Quantum estimation of the Schwarzschild spacetime parameters of the earth. *Phys. Rev. D* **2014**, *90*, 124001. [CrossRef]
27. Alsing, P.M.; Milburn, G.J. Teleportation with a uniformly accelerated partner. *Phys. Rev. Lett.* **2003**, *91*, 180404. [CrossRef] [PubMed]
28. Bruschi, D.E.; Louko, J.; Martín-Martínez, E.; Dragan, A.; Fuentes, I. Unruh effect in quantum information beyond the single-mode approximation. *Phys. Rev. A* **2010**, *82*, 042332. [CrossRef]
29. Bruschi, D.E.; Dragan, A.; Lee, A.R.; Fuentes, I.; Louko, J. Relativistic motion generates quantum gates and entanglement resonances. *Phys. Rev. Lett.* **2013**, *111*, 090504. [CrossRef] [PubMed]
30. Huang, C.Y.; Ma, W.; Wang, D.; Ye, L. How the relativistic motion affect quantum Fisher information and Bell non-locality for multipartite state. *Sci. Rep.* **2017**, *7*, 38456. [CrossRef]
31. Schützhold, R.; Schaller, G.; Habs, D. Signatures of the Unruh Effect from Electrons Accelerated by Ultrastrong Laser Fields. *Phys. Rev. Lett.* **2006**, *97*, 121302. [CrossRef]
32. Schützhold, R.; Schaller, G.; Habs, D. Tabletop Creation of Entangled Multi-keV Photon Pairs and the Unruh Effect. *Phys. Rev. Lett.* **2008**, *100*, 091301.
33. Hawton, M. Photon counting by inertial and accelerated detectors. *Phys. Rev. A* **2013**, *87*, 042116. [CrossRef]
34. Soldati, R.; Specchia, C. On the Massless Vector Fields in a Rindler Space. *J. Mod. Phys.* **2015**, *6*, 1743–1755. [CrossRef]
35. Bennett, R.; Barlow, T.M.; Beige, A. A physically motivated quantization of the electromagnetic field. *Eur. J. Phys.* **2016**, *37*, 014001. [CrossRef]
36. Stokes, A.; Kurcz, A.; Spiller, T.P.; Beige, A. Extending the validity range of quantum optical master equations. *Phys. Rev. A* **2010**, *85*, 053805. [CrossRef]
37. Stokes, A. Noncovariant gauge fixing in the quantum Dirac field theory of atoms and molecules. *Phys. Rev. A* **2012**, *86*, 012511. [CrossRef]
38. Stokes, A.; Deb, P.; Beige, A. Using thermodynamics to identify quantum subsystems. *J. Mod. Opt.* **2017**, *64*, S7–S19. [CrossRef]
39. Parker, L.E.; Toms, D.J. *Quantum Field Theory in Curved Spacetime: Quantized Fields and Gravity*; Cambridge University Press: Cambridge, UK, 2009.

40. Casals, M.; Dolan, S.R.; Nolan, B.C.; Ottewill, A.C.; Winstanley, E. Quantization of fermions on Kerr space-time. *Phys. Rev. D* **2013**, *87*, 064027. [CrossRef]
41. Dirac, P.A.M. Gauge-invariant formulation of quantum electrodynamics. *Can. J. Phys.* **1955**, *33*, 650–660. [CrossRef]
42. DeWitt, B.S. Quantum Theory without Electromagnetic Potentials. *Phys. Rev.* **1962**, *125*, 2189–2191. [CrossRef]
43. Gray, R.D.; Kobe, D.H. Gauge-invariant canonical quantisation of the electromagnetic field and duality transformations. *J. Phys. A Math. Gen.* **1982**, *15*, 3145–3155. [CrossRef]
44. Menikoff, R.; Sharp, D.H. A gauge invariant formulation of quantum electrodynamics using local currents. *J. Math. Phys.* **1977**, *18*, 471–482. [CrossRef]
45. Dirac, P.A.M. *Lectures on Quantum Mechanics*; Dover Publications Inc.: Mineola, NY, USA, 2001.
46. Loudon, R. *The Quantum Theory of Light*; Oxford University Press: Oxford, UK, 2000.
47. Misner, C.W.; Thorne, K.S.; Wheeler, J.A. *Gravitation*; Macmillan: London, UK, 1973.
48. Carroll, S.M. *Spacetime and Geometry: An Introduction to General Relativity*; Addison Wesley: Boston, MA, USA, 2004.
49. Lawrie, I.D. *A Unified Grand Tour of Theoretical Physics*, 2nd ed.; Institute of Physics Publishing: Bristol, UK, 2002.
50. Martín-Martínez, E.; Rodriguez-Lopez, P. Relativistic quantum optics: The relativistic invariance of the light–Matter interaction models. *Phys. Rev. D* **2018**, *97*, 105026. [CrossRef]
51. Birrell, N.D.; Davies, P.C.W. *Quantum Fields in Curved Space*; Cambridge University Press: Cambridge, UK, 1984.
52. Fulling, S.A. Nonuniqueness of canonical field quantization in Riemannian space-time. *Phys. Rev. D* **1973**, *7*, 2850–2862. [CrossRef]
53. Alsing, P.M.; Fuentes, I. Observer-dependent entanglement. *Class. Quantum Gravity* **2012**, *29*, 224001. [CrossRef]
54. Hollands, S.; Wald, R.M. Quantum fields in curved spacetime. *Phys. Rep.* **2015**, *574*, 1–35. [CrossRef]
55. Unruh, W.G. Notes on black-hole evaporation. *Phys. Rev. D* **1976**, *14*, 870–892. [CrossRef]
56. Unruh, W.G.; Weiss, N. Acceleration radiation in interacting field theories. *Phys. Rev. D* **1984**, *29*, 1656–1662. [CrossRef]
57. Takagi, S. Vacuum Noise and Stress Induced by Uniform Acceleration: Hawking-Unruh Effect in Rindler Manifold of Arbitrary Dimension. *Prog. Theor. Phys. Suppl.* **1986**, *88*, 1–142. [CrossRef]
58. Crispino, L.C.B.; Higuchi, A.; Matsas, G.E.A. The Unruh effect and its applications. *Rev. Mod. Phys.* **2008**, *80*, 787–838. [CrossRef]
59. Buchholz, D.; Verch, R. Unruh versus Tolman: On the heat of acceleration. *Gen. Relativ. Gravit.* **2016**, *48*, 1–9. [CrossRef]
60. Acedo, L.; Tung, M.M. Electromagnetic waves in a uniform gravitational field and Planck's postulate. *Eur. J. Phys.* **2012**, *33*, 1073. [CrossRef]
61. de Almeida, C.; Saa, A. The radiation of a uniformly accelerated charge is beyond the horizon: A simple derivation. *Am. J. Phys.* **2006**, *74*, 154–158. [CrossRef]
62. Desloge, E.A. Nonequivalence of a uniformly accelerating reference frame and a frame at rest in a uniform gravitational field. *Am. J. Phys.* **1989**, *57*, 1121–1125. [CrossRef]
63. Rindler, W. Kruskal space and the uniformly accelerated frame. *Am. J. Phys.* **1966**, *34*, 1174–1178. [CrossRef]
64. Semay, C. Observer with a constant proper acceleration. *Eur. J. Phys.* **2006**, *27*, 1157–1167. [CrossRef]
65. Ornigotti, M.; Aiello, A. The Faddeev-Popov Method Demystified. *arXiv* **2014**, arXiv:1407.7256.
66. Andrews, D.L.; Romero, L.C.D. A back-to-front derivation: the equal spacing of quantum levels is a proof of simple harmonic oscillator physics. *Eur. J. Phys.* **2009**, *30*, 1371–1380. [CrossRef]
67. Ballentine, L.E. *Quantum Mechanics: A Modern Development*; World Scientific Publishing Company: Singapore, 1998.
68. Nikolić, H. Horava-Lifshitz gravity, absolute time, and objective particles in curved space. *Mod. Phys. Lett. A* **2010**, *25*, 1595–1601. [CrossRef]
69. Peres, A.; Terno, D.R. Quantum information and relativity theory. *Rev. Mod. Phys.* **2004**, *76*, 93–123. [CrossRef]

70. Brown, J.D.; York, J.W., Jr. Quasilocal energy and conserved charges derived from the gravitational action. *Phys. Rev. D* **1993**, *47*, 1407–1419. [CrossRef]
71. Bogolubov, N.N.; Shirkov, D.V. *Introduction to the Theory of Quantized Fields*; Wiley-Interscience: New York, NY, USA, 1980.
72. Schwinger, J. Quantum Electrodynamics. I. A Covariant Formulation. *Phys. Rev.* **1948**, *74*, 1439–1461. [CrossRef]
73. Schwartz, M.D. *Quantum Field Theory and the Standard Model*; Cambridge University Press: Cambridge, UK, 2014.
74. Kasper, U.; Kreisel, E.; Treder, H.J. On the Covariant Formulation of Quantum Mechanics. *Found. Phys.* **1977**, *7*, 375–389. [CrossRef]
75. Crawford, J.P. Spinor Matter in a gravitational field: Covariant equations àla Heisenberg. *Found. Phys.* **1998**, *28*, 457–470. [CrossRef]
76. Susskind, L. Black Holes and Holography. Available online: https://www.perimeterinstitute.ca/video-library/collection/black-holes-and-holography-mini-course-2007 (accessed on 29 August 2019).
77. Iorio, A.; Lambiase, G.; Vitiello, G. Quantization of Scalar Fields in Curved Background and Quantum Algebras. *Ann. Phys.* **2001**, *294*, 234–250. [CrossRef]

© 2019 by the authors. Licensee MDPI, Basel, Switzerland. This article is an open access article distributed under the terms and conditions of the Creative Commons Attribution (CC BY) license (http://creativecommons.org/licenses/by/4.0/).

Article

Summoning, No-Signalling and Relativistic Bit Commitments

Adrian Kent [1,2]

[1] Centre for Quantum Information and Foundations, DAMTP, Centre for Mathematical Sciences, University of Cambridge, Wilberforce Road, Cambridge CB3 0WA, UK; apak@damtp.cam.ac.uk
[2] Perimeter Institute for Theoretical Physics, 31 Caroline Street North, Waterloo, ON N2L 2Y5, Canada

Received: 22 March 2019; Accepted: 18 May 2019; Published: 25 May 2019

Abstract: Summoning is a task between two parties, Alice and Bob, with distributed networks of agents in space-time. Bob gives Alice a random quantum state, known to him but not her, at some point. She is required to return the state at some later point, belonging to a subset defined by communications received from Bob at other points. Many results about summoning, including the impossibility of unrestricted summoning tasks and the necessary conditions for specific types of summoning tasks to be possible, follow directly from the quantum no-cloning theorem and the relativistic no-superluminal-signalling principle. The impossibility of cloning devices can be derived from the impossibility of superluminal signalling and the projection postulate, together with assumptions about the devices' location-independent functioning. In this qualified sense, known summoning results follow from the causal structure of space-time and the properties of quantum measurements. Bounds on the fidelity of approximate cloning can be similarly derived. Bit commitment protocols and other cryptographic protocols based on the no-summoning theorem can thus be proven secure against some classes of post-quantum but non-signalling adversaries.

Keywords: relativistic quantum information; quantum cryptography; summoning; no-cloning; no-signalling; bit commitment

1. Introduction

To define a summoning task [1,2], we consider two parties, Alice and Bob, who each have networks of collaborating agents occupying non-overlapping secure sites throughout space-time. At some point P, Bob's local agent gives Alice's local agent a state $|\psi\rangle$. The physical form of $|\psi\rangle$ and the dimension of its Hilbert space H are pre-agreed; Bob knows a classical description of $|\psi\rangle$ but from Alice's perspective it is a random state drawn from the uniform distribution on H. At further pre-agreed points (which are often taken to all be in the causal future of P, though this is not necessary), Bob's agents send classical communications in pre-agreed form, satisfying pre-agreed constraints, to Alice's local agents, which collectively determine a set of one or more valid return points. Alice may manipulate and propagate the state as she wishes but must return it to Bob at one of the valid return points. We say a given summoning task is *possible* if there is some algorithm that allows Alice to ensure that the state is returned to a valid return point for any valid set of communications received from Bob.

The "no-summoning theorem" [1] states that summoning tasks in Minkowski space are not always possible. We write $Q \succ P$ if the space-time point Q is in the causal future of the point P and $Q \not\succ P$ otherwise; we write $Q \succeq P$ if either $Q \succ P$ or $Q = P$ and $Q \not\succeq P$ otherwise. Now, for example, consider a task in which Bob may request at one of two "call" points $c_i \succ P$ that the state be returned at a corresponding return point $r_i \succ c_i$, where $r_2 \not\succeq c_1$ and $r_1 \not\succeq c_2$. An algorithm that guarantees that Alice will return the state at r_1 if it is called at c_1 must work independently of whether a call is also made at c_2, since no information can propagate from c_2 to r_1; similarly if 1 and 2 are exchanged. If calls were made at both c_1 and c_2, such an algorithm would thus generate two copies of $|\psi\rangle$ at the space-like separated points r_1 and r_2, violating the no-cloning theorem. This distinguishes relativistic quantum theory from both relativistic classical mechanics and non-relativistic quantum mechanics, in which summoning tasks are always possible provided that any valid return point is in the (causal) future of the start point P.

Further evidence for seeing summoning tasks as characterising fundamental features of relativistic quantum theory was given by Hayden and May [3], who considered tasks in which a request is made at precisely one from a pre-agreed set of call points $\{c_1, \ldots, c_n\}$; a request at c_i requires the state to be produced at the corresponding return point $r_i \succ c_i$. They showed that, if the start point P is in the causal past of all the call points, then the task is possible if and only if no two causal diamonds $D_i = \{x : r_i \succeq x \succeq c_i\}$ are spacelike separated. That is, the task is possible unless the no-cloning and no-superluminal-signalling principles directly imply its impossibility. Wu et al. have presented a more efficient code for this task [4]. Another natural type of summoning task allows any number of calls to be made at call points, requiring that the state be produced at any one of the corresponding return points. Perhaps counter-intuitively, this can be shown to be a strictly harder version of the task [5]. It is possible if and only if the causal diamonds can be ordered in sequence so that the return point of any diamond in the sequence is in the causal future of all call points of earlier diamonds in the sequence. Again, the necessity of this condition follows (with a few extra steps) from the no-superluminal-signalling and no-cloning theorems [5].

The constraints on summoning have cryptographic applications, since they can effectively force Alice to make choices before revealing them to Bob. Perhaps the simplest and most striking of these is a novel type of unconditionally secure relativistic quantum bit commitment protocol, in which Alice sends the unknown state at light speed in one of two directions, depending on her committed bit [6]. The fidelity bounds on approximate quantum cloning imply [6] the sum-binding security condition

$$p_0 + p_1 \leq 1 + \frac{2}{d+1}, \qquad (1)$$

where $d = \dim(H)$ is the dimension of the Hilbert space of the unknown state and p_b is the probability of Alice successfully unveiling bit value b.

Summoning is also a natural primitive in distributed quantum computation, in which algorithms may effectively summon a quantum state produced by a subroutine to some computation node that depends on other computed or incoming data.

From a fundamental perspective, the (im)possibility of various summoning tasks may be seen either as results about relativistic quantum theory or as candidate axioms for a reformulation of that theory. They also give a way of exploring and characterising the space of theories generalising relativistic quantum theory. From a cryptographic perspective, we would like to understand precisely which assumptions are necessary for the security of summoning-based protocols. These motivations are particularly strong given the relationship between no-summoning theorems and no-signalling, since we know that quantum key distribution and other protocols can be proven secure based on no-signalling principles alone. In what follows, we characterise that relationship more precisely, and discuss in particular the sense in which summoning-based bit commitment protocols are secure against potentially post-quantum but non-signalling participants. These are participants who may have access to technology that relies on some unknown theory beyond quantum theory. They may thus be able to carry out operations that quantum theory suggests is impossible. However, their technology must not allow them to violate a no-signalling principle. Exactly what this implies depends on which no-signalling principle is invoked. We turn next to discussing the relevant possibilities.

2. No-Signalling Principles and No-Cloning

2.1. No-Signalling Principles

The relativistic no-superluminal-signalling principle states that no classical or quantum information can be transmitted faster than light speed. We can frame this operationally by considering a general physical system that includes agents at locations $P_1, \ldots, P_n$. Suppose that the agent at each P_i may freely choose inputs labelled by A_i and receive outputs a_i, which may probabilistically depend on their and other inputs. Let $I = \{i_1, \ldots, i_b\}$ and $J = \{j_1, \ldots j_c\}$ be sets of labels of points such that $P_{i_k} \not\succ P_{j_l}$ for all $k \in \{1, \ldots, b\}$ and $l \in \{1, \ldots, c\}$. Then we have

$$P(a_{i_1} \ldots a_{i_b} | A_{i_1} \ldots A_{i_b}) = \tag{2}$$
$$p(a_{i_1} \ldots a_{i_b} | A_{i_1} \ldots A_{i_b} A_{j_1} \ldots A_{j_c}).$$

In other words, outputs are independent of spacelike or future inputs.

The quantum no-signalling principle for an n-partite system composed of non-interacting subsystems states that measurement outcomes on any subset of subsystems are independent of measurement choices on the others. If we label the measurement choices on subsystem i by A_i, and the outcomes for this choice by a_i, then we have

$$P(a_{i_1} \ldots a_{i_m} | A_{i_1} \ldots A_{i_m}) = P(a_{i_1} \ldots a_{i_m} | A_1 \ldots A_n). \tag{3}$$

That is, so long as the subsystems are non-interacting, the outputs for any subset are independent of the inputs for the complementary subset, regardless of their respective locations in space-time.

The no-signalling principle for a generalised non-signalling theory extends this to any notional device with localised pairs of inputs (generalising measurement choices) and outputs (generalising outcomes). As in the quantum case, this is supposed to hold true regardless of whether the sites of the localised input/output ports are spacelike separated. Generalized non-signalling theories may include, for example, the hypothetical bipartite Popescu-Rohrlich boxes [7], which maximally violate the CHSH inequality, while still precluding signalling between agents at each site.

2.2. The No-Cloning Theorem

The standard derivation of the no-cloning theorem [8,9] assumes a hypothetical quantum cloning device. A quantum cloning device D should take two input states, a general quantum state $|\psi\rangle$ and a reference state $|0\rangle$, independent of $|\psi\rangle$. Since D follows the laws of quantum theory, it must act linearly. Now we have

$$D|\psi\rangle|0\rangle = |\psi\rangle|\psi\rangle, \quad D|\psi'\rangle|0\rangle = |\psi'\rangle|\psi'\rangle, \tag{4}$$

for a faithful cloning device, for any states $|\psi\rangle$ and $|\psi'\rangle$. Suppose that $\langle\psi'|\psi\rangle = 0$ and that $|\phi\rangle = a|\psi\rangle + b|\psi'\rangle$ is normalised. We also have

$$D|\phi\rangle|0\rangle = |\phi\rangle|\phi\rangle, \tag{5}$$

which contradicts linearity.

To derive the no-cloning theorem without appealing to linearity, we need to consider quantum theory as embedded within a more general theory that does not necessarily respect linearity. We can then consistently consider a hypothetical post-quantum cloning device D which accepts quantum states $|\psi\rangle$ and $|0\rangle$ as inputs and produces two copies of $|\psi\rangle$ as outputs:

$$D|\psi\rangle|0\rangle = |\psi\rangle|\psi\rangle. \tag{6}$$

We will suppose that the cloning device functions in this way independent of the history of the input state. We will also suppose that it does not violate any other standard physical principles: in particular, if it is applied at Q then it does not act retrocausally to influence the outcomes of measurements at earlier points $P \prec Q$.

We can now extend the cloning device to a bipartite device comprising a maximally entangled quantum state, with a standard quantum measurement device at one end and the cloning device followed by a standard quantum measurement device at the other end. This extended device accepts classical inputs (measurement choices) and produces classical outputs (measurement outcomes) at both ends.

If we now further assume that the joint output probabilities for this extended device, for any set of inputs, are independent of the locations of its components, then we can derive a contradiction with the relativistic no-superluminal signalling principle. First suppose that the two ends are timelike separated, with the cloning device end at point Q and the other end at point $P \prec Q$. A complete projective measurement at P then produces a pure state at Q in any standard version of quantum theory. The cloning device then clones this pure state. Different measurement choices at P produce different ensembles of pure states at Q. These ensembles correspond to the same mixed state before cloning but to distinguishable mixtures after cloning. The measurement device at Q can distinguish these mixtures. Now if we take the first end to be at a point P' spacelike separated from Q, by hypothesis the output probabilities remain unchanged. This allows measurement choices at P' to be distinguished by measurements at Q and so gives superluminal signalling [10].

It is important to note that the assumption of location-independence is not logically necessary, nor does it follow from the relativistic no-superluminal-signalling principle alone. Assuming that quantum states collapse in some well defined and localized way as a result of measurements, one can consistently extend relativistic quantum theory to include hypothetical devices that read out a classical description of the local reduced density matrix at any given point, that is the local quantum state that is obtained by taking into account (only) collapses within the past light cone [11]. This means that measurement events at P, which we take to induce collapses, are taken into account by the readout device at Q if and only if $P \prec Q$. Given such a readout device, one can certainly clone pure quantum states. The device behaves differently, when applied to a subsystem of an entangled system, depending on whether the second subsystem is measured inside or outside the past light cone of the point at which the device is applies. It thus does not satisfy the assumptions of the previous paragraph.

The discussion above also shows that quantum theory augmented by cloning or readout devices is not a generalized non-signalling theory. For consider again a maximally entangled bipartite quantum system with one subsystem at space-time point P and the other at a space-like separated point P'. Suppose that the Hamiltonian is zero and that the subsystem at P' will propagate undisturbed to point $Q \succ P$. Suppose that a measurement device may carry out any complete projective measurement at P and that at Q there is a cloning device followed by another measurement device on the joint (original and cloned) system. As above, different measurement choices at P produce different ensembles of pure states at Q, which correspond to the same mixed state before cloning but to distinguishable mixtures after cloning. The measurement device at Q can distinguish these mixtures. The output (measurement outcome) probabilities at Q thus depend on the inputs (measurement choices) at P, contradicting Equation (3). Assuming that nature is described by a generalized non-signalling theory thus gives another reason for excluding cloning or readout devices, without assuming that their behaviour is location-independent.

In summary, neither the no-cloning theorem nor cryptographic security proofs based on it can be derived purely from consistency with special relativity. They require further assumptions about the behaviour of post-quantum devices available to participants or adversaries. Although this was noted when cryptography based on the no-signalling principle was first introduced [12], it perhaps deserves re-emphasis.

On the positive side, given these further assumptions, one can prove not only the no-cloning theorem but also quantitative bounds on the optimal fidelities attainable by approximate cloning devices for qubits [10] and qudits [13]. In particular, one can show [13] that any approximate universal cloning device that produces output states ρ_0 and ρ_1 given a pure input qudit state $|\psi\rangle$ satisfies the fidelity sum bound

$$\langle\psi|\rho_0|\psi\rangle + \langle\psi|\rho_1|\psi\rangle \leq 1 + \frac{2}{d+1}. \tag{7}$$

It is worth stressing that (with the given assumptions) this bound applies for any approximate cloning strategy, with any entangled states allowed as input.

3. Summoning-Based Bit Commitments and No-Signalling

We recall now the essential idea of the flying qudit bit commitment protocol presented in Reference [6], in its idealized form. We suppose that space-time is Minkowski and that both parties, the committer (Alice) and the recipient (Bob), have arbitrarily efficient technology, limited only by physical principles. In particular, we assume they both can carry out error-free quantum operations instantaneously and can send classical and quantum information at light speed without errors. They agree in advance on some space-time point P, to which they have independent secure access, where the commitment will commence.

We suppose too that Bob can keep a state secure from Alice somewhere in the past of P and arrange to transfer it to her at P. Alice's operations on the state can then be kept secure from Bob unless and until she chooses to return information to Bob at some point(s) in the future of P. We also suppose that Alice can send any relevant states at light speed in prescribed directions along secure quantum channels, either by ordinary physical transmission or by teleportation.

They also agree on a fixed inertial reference frame and two opposite spatial directions within that frame. For simplicity we neglect the y and z coordinates and take the speed of light $c = 1$. Let $P = (0,0)$ be the origin in the coordinates (x,t) and the two opposite spatial directions be defined by the vectors $v_0 = (-1, 0)$ and $v_1 = (1, 0)$.

Before the commitment begins, Bob generates a random pure qudit $|\psi\rangle \in \mathcal{C}^d$. This is chosen from the uniform distribution and encoded in some pre-agreed physical system. Again idealizing, we assume the dimensions of this system are negligible and treat it as pointlike. Bob keeps his qudit secure until the point P, where he gives it to Alice. To commit to the bit $i \in \{0, 1\}$, Alice sends the state $|\psi\rangle$ along a secure channel at light speed in the direction v_i. That is, to commit to 0, she sends the qudit along the line $L_0 = \{(-t, t), t > 0\}$; to commit to 1, she sends it along the line $L_1 = \{(t, t), t > 0\}$.

For simplicity, we suppose here that Alice directly transmits the state along a secure channel. This allows Alice the possibility of unveiling her commitment at any point along the transmitted light ray. To unveil the committed bit 0, Alice returns $|\psi\rangle$ to Bob at some point Q_0 on L_0; to unveil the committed bit 1, Alice returns $|\psi\rangle$ to Bob at some point Q_1 on L_1. Bob then tests that the returned qudit is $|\psi\rangle$ by carrying out the projective measurement defined by $P_\psi = |\psi\rangle\langle\psi|$ and its complement $(I - P_\psi)$. If he gets the outcome corresponding to P_ψ, he accepts the commitment as honestly unveiled; if not, he has detected Alice cheating.

Now, given any strategy of Alice's at P, there is an optimal state ρ_0 she can return to Bob at Q_0 to maximise the chance of passing his test there, i.e., to maximize the fidelity $\langle\psi|\rho_0|\psi\rangle$. There is similarly an optimal state ρ_1 that she can return at Q_1, maximizing $\langle\psi|\rho_1|\psi\rangle$. The relativistic no-superluminal-signalling principle implies that her ability to return ρ_0 at Q_0 cannot depend on whether she chooses to return ρ_1 at Q_1 or vice versa. Hence she may return both (although this violates the protocol). The bound (7) on the approximate cloning fidelities implies that

$$\langle\psi|\rho_0|\psi\rangle + \langle\psi|\rho_1|\psi\rangle \leq 1 + \frac{2}{d+1}. \tag{8}$$

Since the probability of Alice successfully unveiling the bit value b by this strategy is

$$p_b = \langle \psi | \rho_b | \psi \rangle, \qquad (9)$$

this gives the sum-binding security condition for the bit commitment protocol

$$p_0 + p_1 \leq 1 + \frac{2}{d+1}. \qquad (10)$$

Recall that the bound (7) follows from the relativistic no-superluminal-signalling condition together with the location-independence assumption for a device based on a hypothetical post-quantum cloning device applied to one subsystem of a bipartite entangled state. Alternatively, it follows from assuming that any post-quantum devices operate within a generalized non-signalling theory. The bit commitment security thus also follows from either of these assumptions.

3.1. Security Against Post-Quantum No-Superluminal-Signalling Adversaries?

It is a strong assumption that any post-quantum theory should be a generalized non-signalling theory satisfying Equation (3). So it is natural to ask whether cryptographic security can be maintained with the weaker assumption that other participants or adversaries are able to carry out quantum operations and may also be equipped with post-quantum devices but do not have the power to signal superluminally. It is instructive to understand the limitations of this scenario for protocols between mistrustful parties capable of quantum operations, such as the bit commitment protocol just discussed.

The relevant participant here is Alice, who begins with a quantum state at P and may send components along the lightlike lines PQ_0 and PQ_1. Without loss of generality we assume these are the only components: she could also send components in other directions but relativistic no-superluminal-signalling means that they cannot then influence her states at Q_0 or Q_1.

At any points X_0 and X_1 on the lightlike lines, before Alice has applied any post-quantum devices, the approximate cloning fidelity bound again implies that fidelities of the respective components ρ_{X_0} and ρ_{X_1} satisfy

$$\langle \psi | \rho_{X_0} | \psi \rangle + \langle \psi | \rho_{X_1} | \psi \rangle \leq 1 + \frac{2}{d+1}. \qquad (11)$$

Now, if Alice possesses a classical no-superluminal-signalling device, such as a Popescu-Rohrlich box, with input and output ports at X_0 and X_1 and her agents at these sites input classical information uncorrelated with their quantum states, she does not alter the fidelities $\langle \psi | \rho_{X_i} | \psi \rangle$. Any subsequent operation may reduce the fidelities but cannot increase them. More generally, any operation involving the quantum states and devices with purely classical inputs and outputs cannot increase the fidelity sum bound (7). To see this, note that any such operation could be paralleled by local operations within quantum theory if the two states were held at the same point, since hypothetical classical devices with separated pairs of input and output ports are replicable by ordinary probabilistic classical devices when the ports are all at the same site.

We need also to consider the possibility that Alice has no-superluminal signalling devices with quantum inputs and outputs. At first sight these may seem unthreatening. For example, while a device that sends the quantum input from X_0 to the output at X_1 and vice versa would certainly make the protocol insecure—Alice could freely swap commitments to 0 and 1—such a device would be signalling.

However, suppose that Alice's agents each have local state readout devices, which give Alice's agent at X_0 a classical description of the density matrix ρ_{X_0} and Alice's agent at X_1 a classical description of the density matrix ρ_{X_1}. Suppose also that Alice has carried out an approximate universal cloning at P, creating mixed states ρ_{X_0} and ρ_{X_1} of the form

$$\rho_{X_i} = p_i | \psi \rangle \langle \psi | + (1 - p_i) I, \qquad (12)$$

where $0 < p_i < 1$. This is possible provided that $p_0 + p_1 \leq 1 + \frac{2}{d+1}$. From these, by applying their readout devices, each agent can infer $|\psi\rangle$ locally. Alice's outputs at X_i have no dependence on the inputs at $X_{\bar{i}}$. Nonetheless, this hypothetical process would violate the security of the commitment to the maximum extent possible, since it would give $p_0 + p_1 = 2$.

To ensure post-quantum security, our post-quantum theory thus need assumptions—like those spelled out earlier—that directly preclude state readout devices and other violations of no-cloning bounds.

4. Discussion

Classical and quantum relativistic bit commitment protocols have attracted much interest lately, both because of their theoretical interest and because advances in theory [14] and practical implementation [15–17] suggest that relativistic cryptography may be in widespread use in the forseeable future.

Much work on these topics is framed in models in which two (or more) provers communicate with one (or more) verifiers, with the provers being unable to communicate with one another during the protocol. Indeed, one round classical relativistic bit commitment protocols give a natural physical setting in which two (or more) separated provers communicate with adjacent verifiers, with the communications timed so that the provers cannot communicate between the commitment and opening phases. The verifiers are also typically unable to communicate but this is less significant given the form of the protocols and the verifiers are sometimes considered as a single entity when the protocol is not explicitly relativistic.

Within the prover-verifier model, it has been shown that no single-round two-prover classical bit commitment protocol can be secure against post-quantum provers who are equipped with generalized no-signalling devices [18]. It is interesting to compare this result with the signalling-based security proof for the protocol discussed above.

First, of course, the flying qudit protocol involves quantum rather than classical communication between "provers" (Alice's agents) and "verifiers" (Bob's agents).

Second, as presented, the flying qudit protocol involves three agents for each party. However, a similar secure bit commitment protocol can be defined using just two agents apiece. For example, Alice's agent at P could retain the qudit, while remaining stationary in the given frame, to commit to 0, and send it to Alice's agent at Q_1 (as before) to commit to 1. They may unveil by returning the qudit at, respectively, $(0,t)$ or (t,t). In this variant, the commitment is not secure at the point where the qudit is received, but it becomes secure in the causal future of $(t/2, t/2)$.

Third, the original flying qudit protocol illustrates a possibility in relativistic quantum cryptography that is not motivated (and so not normally considered) in standard multi-prover bit commitment protocols. This is that, while there are three provers, communication between them in some directions is possible (and required) during the protocol. Alice's agent at P must be able to send the quantum state to either of the agents at Q_0 or Q_1; indeed, a general quantum strategy requires her to send quantum information to both.

Fourth, the security proof of the flying qudit protocol can be extended to generalised no-signalling theories. However, the protocol is not secure if the committer may have post-quantum devices that respect the no-superluminal signalling principle but are otherwise unrestricted. Security proofs require stronger assumptions, such as that the commmitter is restricted to devices allowed by a generalized non-signalling theory.

The same issue arises considering the post-quantum security of quantum key distribution protocols [12], which are secure if a post-quantum eavesdropper is restricted by a generalised no-signalling theory but not if she is only restricted by the no-superluminal-signalling principle. One distinction is that quantum key distribution is a protocol between mutually trusting parties, Alice and Bob, whereas bit commitment protocols involve two mistrustful parties. It is true that quantum key distribution still involves mistrust, in that Alice and Bob mistrust the eavesdropper,

Eve. However, if one makes the standard cryptographic assumption that Alice's and Bob's laboratories are secure, so that information about operations within them cannot propagate to Eve, one can justify a stronger no-signalling principle [12]. Of course, the strength of this justification may be questioned, given that one is postulating unknown physics that could imply a form of light speed signalling that cannot be blocked. But in any case, the justification is not available when one considers protocols between two mistrustful parties, such as bit commitment, and wants to exclude the possibility that one party (in our case Alice) cannot exploit post-quantum operations within her own laboratories (which may be connected, forming a single extended laboratory).

Our discussion assumed a background Minkowski space-time but generalizes to other space-times with standard causal structure, where the causal relation $\prec$ is a partial ordering. Neither standard quantum theory nor the usual form of the no-superluminal signalling principle hold in space-times with closed time-like curves, where two distinct points P and Q may obey both $P \prec Q$ and $Q \prec P$. Formulating consistent theories in this context requires further assumptions (see for example Reference [19] for one analysis). The same is true of superpositions of space-times with indefinite causal order [20]. We leave investigation of these cases for future work.

Funding: This research was funded by UK Quantum Communications Hub grant no. EP/M013472/1 and by the Perimeter Institute for Theoretical Physics. Research at the Perimeter Institute is supported by the Government of Canada through Industry Canada and by the Province of Ontario through the Ministry of Research and Innovation.

Acknowledgments: I thank Claude Crépeau and Serge Fehr for stimulating discussions and the Bellairs Research Institute for hospitality.

Conflicts of Interest: The author declares no conflict of interest.

References

1. Kent, A. A no-summoning theorem in relativistic quantum theory. *Quantum Inf. Process.* **2013**, *12*, 1023–1032. [CrossRef]
2. Kent, A. Quantum tasks in Minkowski space. *Class. Quantum Grav.* **2012**, *29*, 224013. [CrossRef]
3. Hayden, P.; May, A. Summoning information in spacetime, or where and when can a qubit be? *J. Phys. A Math. Theor.* **2016**, *49*, 175304. [CrossRef]
4. Wu, Y.; Khalid, A.; Sanders, B. Efficient code for relativistic quantum summoning. *New J. Phys.* **2018**, *20*, 063052. [CrossRef]
5. Adlam, E.; Kent, A. Quantum paradox of choice: More freedom makes summoning a quantum state harder. *Phys. Rev. A* **2016**, *93*, 062327. [CrossRef]
6. Kent, A. Unconditionally secure bit commitment with flying qudits. *New J. Phys.* **2011**, *13*, 113015. [CrossRef]
7. Popescu, S.; Rohrlich, D. Quantum nonlocality as an axiom. *Found. Phys.* **1994**, *24*, 379–385. [CrossRef]
8. Wootters, W.K.; Zurek, W.H. A single quantum cannot be cloned. *Nature* **1982**, *299*, 802–803. [CrossRef]
9. Dieks, D.G.B.J. Communication by EPR devices. *Phys. Lett. A* **1982**, *92*, 271–272. [CrossRef]
10. Gisin, N. Quantum cloning without signaling. *Phys. Lett. A* **1998**, *242*, 1–3. [CrossRef]
11. Kent, A. Nonlinearity without superluminality. *Phys. Rev. A* **2005**, *72*, 012108. [CrossRef]
12. Barrett, J.; Hardy, L.; Kent, A. No signaling and quantum key distribution. *Phys. Rev. Lett.* **2005**, *95*, 010503. [CrossRef] [PubMed]
13. Navez, P.; Cerf, N.J. Cloning a real d-dimensional quantum state on the edge of the no-signaling condition. *Phys. Rev. A* **2003**, *68*, 032313. [CrossRef]
14. Chailloux, A.; Leverrier, A. Relativistic (or 2-prover 1-round) zero-knowledge protocol for NP secure against quantum adversaries. In Proceedings of the Annual International Conference on the Theory and Applications of Cryptographic Techniques, Paris, France, 30 April–4 May 2017; Springer: Berlin/Heidelberg, Germany, 2017; pp. 369–396.
15. Lunghi, T.; Kaniewski, J.; Bussieres, F.; Houlmann, R.; Tomamichel, M.; Kent, A.; Gisin, N.; Wehner, S.; Zbinden, H. Experimental bit commitment based on quantum communication and special relativity. *Phys. Rev. Lett.* **2013**, *111*, 180504. [CrossRef] [PubMed]

16. Liu, Y.; Cao, Y.; Curty, M.; Liao, S.-K.; Wang, J.; Cui, K.; Li, Y.-H.; Lin, Z.-H.; Sun, Q.-C.; Li, D.-D.; et al. Experimental unconditionally secure bit commitment. *Phys. Rev. Lett.* **2014**, *112*, 010504. [CrossRef] [PubMed]
17. Verbanis, E.; Martin, A.; Houlmann, R.; Boso, G.; Bussières, F.; Zbinden, H. 24-h relativistic bit commitment. *Phys. Rev. Lett.* **2016**, *117*, 140506. [CrossRef] [PubMed]
18. Fehr, S.; Fillinger, M. Multi-prover commitments against non-signaling attacks. In Proceedings of the Annual Cryptology Conference, Santa Barbara, CA, USA, 16–20 August 2015; Springer: Berlin/Heidelberg, Germany, 2015; pp. 403–421.
19. Bennett, C.H.; Leung, D.; Smith, G.; Smolin, J.A. Can closed timelike curves or nonlinear quantum mechanics improve quantum state discrimination or help solve hard problems? *Phys. Rev. Lett.* **2009**, *103*, 170502. [CrossRef] [PubMed]
20. Oreshkov, O.; Costa, F.; Brukner, Č. Quantum correlations with no causal order. *Nat. Commun.* **2012**, *3*, 1092. [CrossRef] [PubMed]

© 2019 by the author. Licensee MDPI, Basel, Switzerland. This article is an open access article distributed under the terms and conditions of the Creative Commons Attribution (CC BY) license (http://creativecommons.org/licenses/by/4.0/).

Article

Simultaneous Classical Communication and Quantum Key Distribution Based on Plug-and-Play Configuration with an Optical Amplifier

Xiaodong Wu [1], Yijun Wang [1,*], Qin Liao [1], Hai Zhong [1] and Ying Guo [1,2,*]

1. School of Automation, Central South University, Changsha 410083, China; wuxiaodong2019@163.com (X.W.); llqqlq@csu.edu.cn (Q.L.); zhonghai@csu.edu.cn (H.Z.)
2. Jiangsu Key Construction Laboratory of IoT Application Technology, Wuxi Taihu University, Wuxi 214064, China
* Correspondence: csuyijun@163.com (Y.W.); yingguo@csu.edu.cn (Y.G.)

Received: 28 January 2019; Accepted: 26 March 2019; Published: 27 March 2019

Abstract: We propose a simultaneous classical communication and quantum key distribution (SCCQ) protocol based on plug-and-play configuration with an optical amplifier. Such a protocol could be attractive in practice since the single plug-and-play system is taken advantage of for multiple purposes. The plug-and-play scheme waives the necessity of using two independent frequency-locked laser sources to perform coherent detection, thus the phase noise existing in our protocol is small which can be tolerated by the SCCQ protocol. To further improve its capabilities, we place an optical amplifier inside Alice's apparatus. Simulation results show that the modified protocol can well improve the secret key rate compared with the original protocol whether in asymptotic limit or finite-size regime.

Keywords: simultaneous; classical communication; quantum key distribution; plug-and-play configuration; optical amplifier

1. Introduction

Quantum key distribution (QKD) is one of the most active areas in quantum information science, which promises to generate a secure key between two authenticated parties (Alice and Bob) over insecure quantum and classical channels [1–4]. The security of a key is guaranteed by the fundamental laws of quantum mechanics [5,6]. Generally speaking, there are two main approaches for the implementation of QKD, namely, discrete-variable (DV) QKD [7] and continuous-variable (CV) QKD [8–11]. Different from the DVQKD, in CVQKD, there is no requirement to use expensive single-photon detectors. Instead, the key bits are encoded in the quadrature variables (X and P) of the optical field, and the secret key bits are decoded through high-efficiency homodyne or heterodyne detection techniques [12–15].

At present, the CVQKD protocol, especially for the Gaussian-modulated coherent-state (GMCS) scheme, has been demonstrated over 100-km telecom fiber through controlling excess noise [16] and designing high-efficiency error correction codes [17–19]. From a practical point of view, the hardware needed in the implementation of the GMCS QKD is amazingly similar to that needed in classical coherent optical communication [20]. On the basis of this similarity, it is viable to take advantage of the same communication facility for both QKD and classical communication.

Recently, a simultaneous classical communication and quantum key distribution (SCCQ) scheme was proposed [21,22]. In this scheme, the Gaussian distributed random numbers for GMCS QKD and bits for classical communication are encoded on the same weak coherent pulse and decoded through the same coherent receiver, which provides a more cost-effective solution in practice. However, a major obstacle to the SCCQ scheme is that it can only tolerate a very small amount of phase noise [21]. This problem could lead to poor performance and thus obstruct its further development.

In this paper, we propose a SCCQ protocol based on plug-and-play configuration with an optical amplifier. Different from the GMCS QKD protocols using a true local oscillator (LO) [23–25], the plug-and-play CVQKD scheme waives the necessity of using two independent frequency-locked laser sources and automatically compensating the polarization drifts [26]. Since the LO and signal pulses are generated from the same laser source, we can obtain a small phase noise which can be tolerated by the SCCQ protocol in the framework of plug-and-play configuration. To further improve its capabilities, we insert an optical amplifier (OA) at the output of the quantum channel [27–29]. The modified protocol (SCCQ protocol based on plug-and-play configuration with an OA) can well increase the secret key rate by compensating the imperfection of Alice's detector with only little cost in transmission distance. Here, both the asymptotic limit and the finite-size regime are taken into consideration.

This paper is structured as follows. In Section 2, we first introduce the plug-and-play dual-phase modulated coherent states (DPMCS) scheme, then present the model of SCCQ protocol based on plug-and-play configuration and the proposed modified protocol. In Section 3, we perform the noise analysis and show numeric simulations in view of practical system parameters. Finally, conclusions are drawn in Section 4. Detailed calculation of equations is shown in the Appendix.

2. Protocol Description

In this section, we first introduce the plug-and-play DPMCS protocol. Then, we present the model of SCCQ protocol based on the plug-and-play configuration and its modified protocol (with OA). To simplify the analysis, we adopt the binary phase-shift keying (BPSK) modulation for the classical communication and the GMCS for QKD protocol in this paper.

2.1. The Plug-and-Play DPMCS Protocol

The prepare-and-measurement scheme of plug-and-play DPMCS protocol is illustrated in Figure 1. The source of light is sent from Alice to Bob, then Bob performs the dual-phase-modulation work after receiving the light. During the modulated process, random numbers drawn from a random number generator are utilized to modulate the amplitude and phase quadrature (X and P quadrature). This is really different from previous one-way CV-QKD protocols where the symmetrical Gaussian modulation is performed at Alice's side. When Bob completes the modulation work, the dual-quadrature modulated coherent-state is directly reflected to Alice with the help of Faraday mirrors. After passing through the untrusted channel characterized by transmittance T and excess noise ξ, Alice receives the modulated signal. Then, she performs homodyne detection to measure the incoming mode. After this, Alice can obtain the list of data which is correlated with the list of Bob. Note that this correction is important in generating a secret key through error reconciliation and privacy amplification. Here, the classical source mentioned above is controlled by Fred [26]. Besides, the untrusted source noise is characterized by taking advantage of a phase-insensitive amplifier (PIA) with a gain of g. In such a practical scheme, the detector used by Alice features an electronic noise v_{el} and an efficiency η. Therefore, the detector-added noise referred to Alice's input can be expressed as $\chi_{hom} = [(1-\eta) + v_{el}]/\eta$.

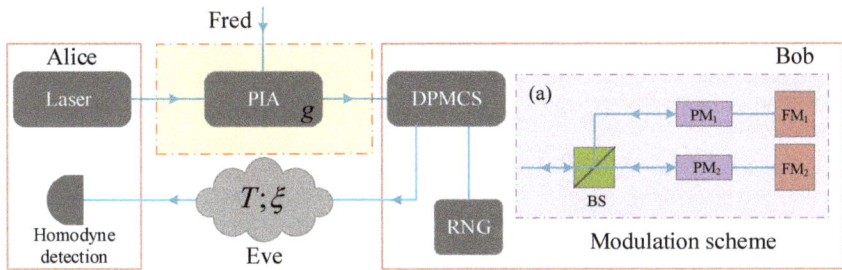

Figure 1. The prepared-and-measure scheme of plug-and-play dual-phase modulated coherent states (DPMCS) protocol. (a) Gaussian modulation scheme by using two phase modulators. PIA, phase insensitive amplifier; RNG, random number generator; PM, phase modulator; FM, Faraday mirror.

2.2. SCCQ Protocol Based on Plug-and-Play Configuration

In the BPSK modulation scheme, the bit value k_B is encoded by $|e^{-ik_B\pi}\alpha\rangle$, where α is a real number. While, in plug-and-play DPMCS protocol, Bob prepares coherent state $|x_B + ip_B\rangle$ and transmits it to Alice. Here x_B and p_B are assumed to be Gaussian random numbers with zero mean and a variance of $V_B N_0$, where N_0 represents the shot-noise variance. The SCCQ protocol based on plug-and-play configuration is straightforward and combines these two communication schemes. Namely, as shown in Figure 2, both the classical bit k_B and Gaussian random numbers $\{x_B, p_B\}$ are encoded on a coherent state $|(x_B + e^{-ik_B\pi}\alpha) + i(p_B + e^{-ik_B\pi}\alpha)\rangle$. It is remarkable that in the plug-and-play DPMCS protocol, Alice performs homodyne detection to measure either the X or P quadrature of each incoming signal. In order to obtain deterministic classical communication, the same classical bit k_B should be encoded on both X and P quadratures.

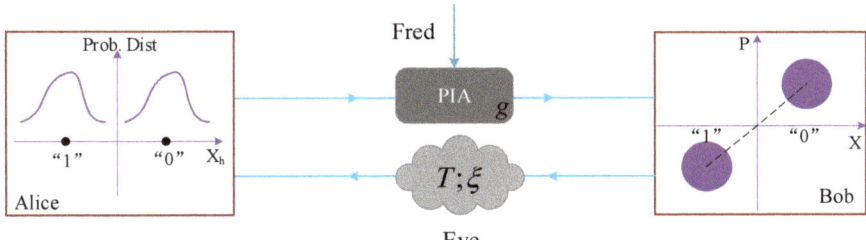

Figure 2. Simultaneous classical communication and quantum key distribution (SCCQ) protocol based on plug-and-play configuration. The probability distributions of X-quadrature measurement is shown at Alice's side.

Suppose Alice measures the X quadrature (P quadrature) and obtains the measurement result x_h (p_h). The sign of x_h (p_h) can be utilized to determine a classical bit k_B. In other words, the value of k_B is assigned as 0 if x_h (p_h) > 0 and the value of k_B is assigned as 1 if x_h (p_h) < 0. Note that according to the overall transmittance $T\eta$ and the value of k_B, Alice's measurement result can be rescaled and displaced to generate a secure key, which is given by

$$\begin{aligned} x_A &= \frac{x_h}{\sqrt{T\eta}} + (2k_B - 1)\alpha, \\ p_A &= \frac{p_h}{\sqrt{T\eta}} + (2k_B - 1)\alpha. \end{aligned} \quad (1)$$

On the basis of the raw keys $\{x_B, x_A\}$ and $\{p_B, p_A\}$, Alice and Bob can distill a secure key by proceeding with classical data postprocessing, as in the case of traditional GMCS QKD.

The prepare-and-measurement (PM) version of our protocol shown above is equivalent to the entanglement-based (EB) version. In the EB scheme, Fred prepares a three-mode entanglement state $|\Phi_{ABF}\rangle$. Bob keeps one mode (B) with variance $V = V_B + 1$ and measures it by using a heterodyne detector. The other mode (A_0) is sent to Alice through an untrusted quantum channel. At Alice's side, a beam splitter with transmission η is taken advantage of to model her detector inefficiency, while an EPR state of variance v_{el} is utilized to model its electronics. For the homodyne detection case, we have $v_{el} = \eta \chi_{hom}/(1-\eta) = 1 + v_{el}/(1-\eta)$. Finally, to distill the secret key, Alice and Bob perform information reconciliation and privacy amplification procedures. Here, we mainly consider the reverse reconciliation since it has been proved to provide a great advantage in performance of QKD schemes [11].

2.3. Addition of an Optical Amplifier

In practice, becuase of some inherent imperfection inevitably existing in Alice's detection apparatus, the ideal detection process cannot be achieved. Therefore, we can only obtain a lower secret key rate than expected. In order to improve the performance of our protocol, here, an optical amplifier is applied to compensate for the detectors' imperfections. In the following, two types of amplifiers are considered, namely, the phase-sensitive amplifier (PSA) and phase-insensitive amplifier (PIA).

Phase-sensitive amplifier. The PSA can be deemed as a degenerate amplifier which allows ideally noiseless amplification of a chosen quadrature. We use a matrix Ξ^{PSA} to describe its properties, which is given by

$$\Xi^{PSA} = \begin{pmatrix} \sqrt{G} & 0 \\ 0 & \sqrt{\frac{1}{G}} \end{pmatrix}, \quad (2)$$

where G represents the gain of amplification and $G \geq 1$.

Phase-insensitive amplifier. The PIA can be regarded as a non-degenerate amplifier, which is able to amplify both quadratures symmetrically. Different from the PSA, the amplification process of the PIA is related to the inherent noise. The transform of the PIA can be modeled as

$$\Xi^{PIA} = \begin{pmatrix} \sqrt{g} I_2 & \sqrt{g-1} \sigma_z \\ \sqrt{g-1} \sigma_z & \sqrt{g} I_2 \end{pmatrix}. \quad (3)$$

The inherent noise of the PIA can be given by

$$\Xi_{noise} = \begin{pmatrix} N_s I_2 & \sqrt{N_s^2 - 1} \sigma_z \\ \sqrt{N_s^2 - 1} \sigma_z & N_s I_2 \end{pmatrix}, \quad (4)$$

where g is the gain of the PIA and N_s stands for variance of noise. We have introduced the PIA in the above analysis. Different from the PIA which is inserted into the output of the quantum channel in our protocol, the PIA is placed at the channel to characterize the untrusted source noise. That is to say the gain g of the PIA can be used to weight the source noise in the plug-and-play scheme.

As illustrated in Figure 3, after the amplification process, mode A_3 is measured using Alice's detector. A beam splitter with transmission η is taken advantage of to model her detector inefficiency. Besides, an EPR state of variance v_{el} is utilized to model its electronics. It is worth mentioning that we adopt homodyne detection in our scheme, thus it is suitable for us to choose the PSA to compensate for Alice's apparatus imperfection [27,28]. Then, the modified parameter χ_{hom}^{PSA} for this case is given by

$$\chi_{hom}^{PSA} = \frac{(1-\eta) + v_{el}}{G \eta}. \quad (5)$$

Consequently, we can achieve the modified secure key rate $\hat{K}$ by substituting χ_{hom}^{PSA} for χ_{hom} in homodyne detection case.

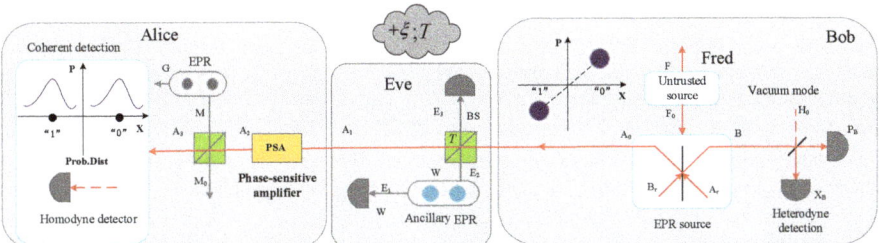

Figure 3. Schematic diagram of the modified protocol (SCCQ protocol based on plug-and-play configuration with an optical amplifier).

3. Performance Analysis and Discussion

The noises which originated from the practical system have important effects on the performance of the SCCQ protocol. In this section, we first introduce the noise model which we adopt in this paper and present the computation of the BER in BPSK modulation scheme. Then, we show and discuss the simulation results.

3.1. Noise Model of SCCQ Protocol Based on Plug-and-Play Configuration

Note that the main noise sources analyzed here are (1) the detector noise assumed as v_{el}, (2) the vacuum noise, (3) the excess noise due to the untrusted sources denoted by ζ_s, (4) excess noise ξ_{RB} caused by Rayleigh backscattering photons, (5) the Gaussian modulation for QKD with a variance of V_B. All the noises mentioned above are defined in the shot-noise unit.

Now let's calculate the BER of the BPSK modulation scheme, which is expressed by [21]

$$BER = \frac{1}{2}erfc(\frac{\sqrt{T\eta}\alpha}{\sqrt{2N_0(V_B T\eta + v_{el} + 1)}}), \quad (6)$$

where $erfc(x)$ represents the complementary error function. In order to make the value of BER small enough in the classical channel, namely, obtain a BER of 10^{-9}, the displacement α is required as

$$\alpha = \frac{4.24\sqrt{V_B T\eta + v_{el} + 1}}{\sqrt{2T\eta}}. \quad (7)$$

The numerical simulations of the required displacement α as a function of the transmission distance and modulation variance V_B are illustrated in Figure 4. It shows that the longer transmission distance needs a larger displacement α for a typical modulation variance V_B in the range of 1 to 20.

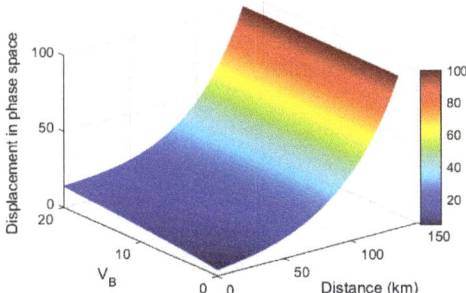

Figure 4. The required displacement α as a function of modulation variance V_B and transmission distance to obtain a BER of 10^{-9} in the classical channel. Parameters $\gamma = 0.2dB/km$, $\eta = 0.5$, and $v_{el} = 0.1$.

The untrusted source noise ζ_s is deemed to be one of the most important excess noises in the plug-and-play configuration. It can be expressed as $\zeta_s = (g-1) + (g-1)V_I$, where g is a gain of a PIA and V_I is the noise variance of a vacuum state (X_I, P_I). That is to say, the untrusted source noise ζ_s can be weighted by parameter g.

The other excess noise we need to consider here is ζ_{RB}, which is caused by Rayleigh backscattering photons. Since the reflected light in the plug-and-play configuration is of the same frequency as the initial laser source, we cannot use the "in-band" photon to filter or attenuate it. The excess noise ζ_{RB} is given by

$$\zeta_{RB} = \frac{2\langle N_{RB}\rangle}{\eta T}, \tag{8}$$

where $\langle N_{RB}\rangle$ is Rayleigh backscattering photons. Then, the backscattered photons $\langle N_{RB}\rangle$ per second Δ_B is expressed as [30]

$$\Delta_B = \frac{\varpi(1 - 10^{-2\gamma L/10})V_B R}{2\eta_B T}, \tag{9}$$

where ϖ stands for the the Rayleigh backscattering coefficient, η_B represents the insertion loss inside Bob (round-trip), R represents the system repetition rate, γ is a fiber loss, and L is the length of an infinite fiber used as a QKD link. Under the assumption that the electronic integral time of Alice's homodyne detector is σt, the excess noise ζ_{RB} can be rewritten as

$$\zeta_{RB} = \frac{2\Delta_B \eta \sigma t}{\eta_B T} = \frac{\varpi(1 - 10^{-2\gamma L/10})V_B R\sigma t}{\eta_B 10^{-2\gamma L/10}}. \tag{10}$$

Note that Equation (10) shows the excess noise which is caused by the quantum channel, namely, here ζ_{RB} can be used to represent ζ.

In the following, we perform an analysis of the effect of phase noise, which commonly exists in a coherent communication system. The excess noise caused by the phase instability is given by

$$\zeta_p = \frac{\alpha^2 \varphi}{N_0}, \tag{11}$$

where φ represents the phase-noise variance. Here Equation (11) is derived with the assumption of $\alpha^2 \geq (V_B + 1)N_0$ [21]. It is worth mentioning that the excess noise φ not only contains the phase noise between the signal and the LO but also the other modulation errors.

On the basis of the above analysis, the overall excess noise outside Alice's system can be defined as

$$\zeta_t = \zeta_s + \zeta_{RB} + \zeta_p. \tag{12}$$

Note that excess noises ζ_s and ζ_{RB} are independent of α.

3.2. Simulation Results

In Figure 5, we conduct numerical simulations of the asymptotic secret key rate as a function of transmission distance in different imperfect source scenarios. Note that $g = 1$ means no source noise case. We adopt the optimal value of modulation V_B in the analysis (see Appendix A). Here, the solid lines in Figure 5 stand for the case of the original protocol ($G = 1$), while the dashed lines represent the case of the modified protocol (a protocol with homodyne detection and a PSA, G = 3). On the one hand, we observed that for each imperfect source scenario, the secret key rate is well improved within a relatively long distance by utilizing an optical preamplifier. On the other hand, we also found that the maximum secure distance of the modified protocol is slightly shorter compared with the original protocol. That is to say, by utilizing the optical amplifier, the secret key rate of the modified protocol increases in a large range of distance with a slight cost of the maximum transmission distance. It is remarkable that the PLOB bound has been plotted in Figure 5, which illustrates the ultimate limit of point-to-point QKD [31]. Here, we should note that the phase noise in our protocol is very small since the LO and signal pulses are generated from the same laser source in the plug-and-play scheme. Therefore, we can achieve the phase noise $\zeta_p = 10^{-6}$. Detailed calculation of the asymptotic secret key rate is shown in Appendix B.

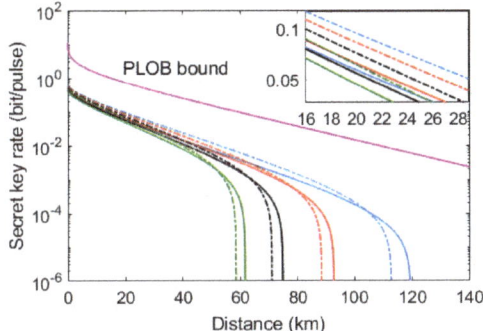

Figure 5. Comparison of secret key rate between the modified protocol (homodyne detection with phase-sensitive amplifier (PSA)) and the original protocol (without PSA) under different imperfect source scenarios. Solid lines represent the original protocol (G = 1) while the dashed lines represent the modified protocol (G = 3). From left to right, the green curves correspond to $g = 1.015$, the black curves correspond to $g = 1.01$, the red curves correspond to $g = 1.005$, and the blue curves correspond to $g = 1$ (no source noise). The simulation parameters are $V_B = 4$, $\zeta_p = 10^{-6}$, $\zeta_{RB} = 0.02$, $\eta = 0.5$, $v_{el} = 0.1$.

In addition, it is necessary to consider the finite-size effect since the length of secret key is impossibly unlimited in practice. Different from the asymptotic case, in the finite-size scenario, the characteristics of the quantum channel cannot be known before the transmission is performed. The reason is that a portion of the exchanged signals needs to be taken advantage of for parameter estimation instead of generating the secret key. We conduct numerical simulations of the finite-size secret key rate in different imperfect source scenarios, as shown in Figure 6. The solid lines in Figure 6 stand for the case of the original protocol ($G = 1$), while the dashed lines represent the case of the modified protocol ($G = 3$). From left to right, the green curves, the black curves, and the red curves correspond to the finite-size scenario of block length $N = 10^6$, 10^8, and 10^{10}, respectively, and the blue curves represent the asymptotic scenario. Here, Figure 6a–d show the proposed protocol with $g = 1$ (no source noise), $g = 1.005$, $g = 1.01$, and $g = 1.015$. We observe that the performance of the asymptotic scenario is better than that of the finite-size scenario whether the PSA is placed at Alice's detection apparatus or not. Furthermore, the curves of the finite-size scenario are more and more close to the curve of asymptotic case with the increased number of exchanged signals N. That is to say that the more exchanged signals we have, the more the signal parameter estimation step can be utilized, and thus the parameter estimation is approaching perfection. Interestingly, for each imperfect source scenario, the finite-size secret key rate of the modified protocol is well improved without the price of reducing the maximum transmission distance, especially for the small-length block, compared with the original protocol, which is different from the asymptotic case. Detailed calculation of finite-size secret key rate is shown in Appendix C.

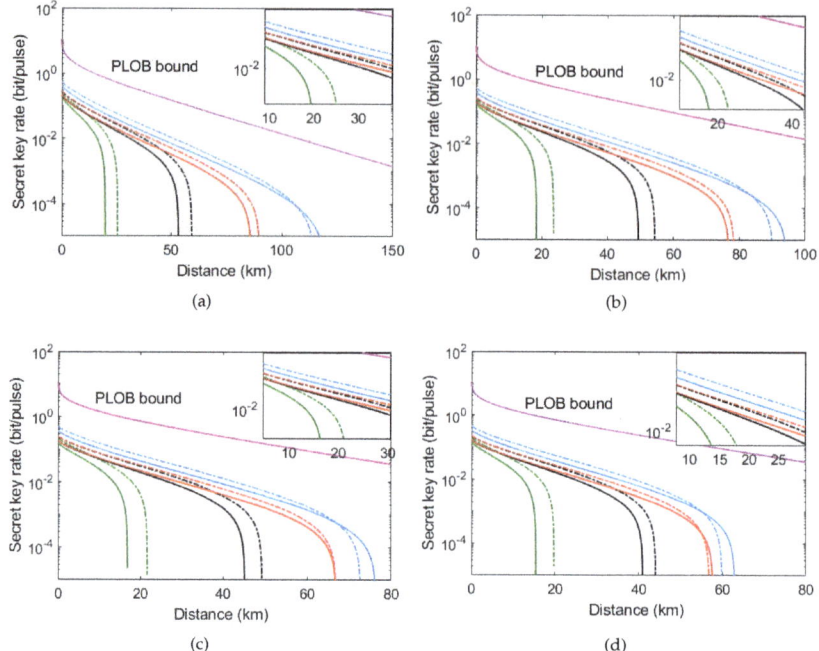

Figure 6. Finite-size secret key rate of SCCQ protocol based on plug-and-play configuration with PSA as a function of transmission distance under different imperfect source scenarios. Solid lines represent the original protocol (G = 1) while the dashed lines represent the modified protocol (G = 3). From left to right, the green curves, the black curves, and the red curves correspond to finite-size scenario of block length $N = 10^6$, 10^8, and 10^{10}, respectively, and the blue curves represent the asymptotic scenario. (**a**) The parameter $g = 1$ (no source noise). (**b**) The parameter $g = 1.005$. (**c**) The parameter $g = 1.01$. (**d**) The parameter $g = 1.015$. Other parameters are set to be the same as Figure 5.

4. Conclusions

We propose a SCCQ protocol based on plug-and-play configuration with an optical amplifier. Benefiting from the plug-and-play scheme where a real local LO is generated from the same laser of quantum signal at Alice's side, the phase noise existing in our protocol is very small, which can be tolerated by the SCCQ protocol. Therefore, our research may bring the SCCQ technology into real life and thus reduce the cost of QKD effectively. To further improve its capabilities, we inserted an optical amplifier inside Alice's apparatus. The simulation results show that the secret key rate is greatly enhanced in a large range of distances for each imperfect source scenario in both asymptotic limit and finite-size regime compared with the original protocol.

Author Contributions: Y.W. and Y.G. gave the general idea of the study, designed the conception of the study and performed critical revision of the manuscript. X.W. accomplished the formula derivation and numerical simulations and drafted the article. Q.L. provided feasible advices and critical revision of the manuscript. All authors have read and approved the final manuscript.

Funding: This work was supported by the National Natural Science Foundation of China (Grant Nos. 61572529, 61871407, 61872390, 61801522), and the Natural Science Foundation of the Jiangsu Higher Education Institutions of China (Grant No. 18KJB510045).

Conflicts of Interest: The authors declare no conflict of interest.

Appendix A. Parameter Optimization

To maximize the performance of our protocol, we need to find an optimal Bob's modulation V_B. As illustrated in Figures A1–A3, different values of g, d, and G are, respectively, set to find a public optimal V_B. From Figure A1, we observe that as the value of parameter g increases (the source noise increases), the optimal interval becomes gradually compressed. In addition, the secret key rate also decreases as a result of the increase in parameter g. Fortunately, there exists a public interval where we can obtain a public optimal modulation V_B for all curves in Figure A1. Namely, we have $V_B = 4$. Note that in this case, the parameters d and G are fixed to legitimate values.

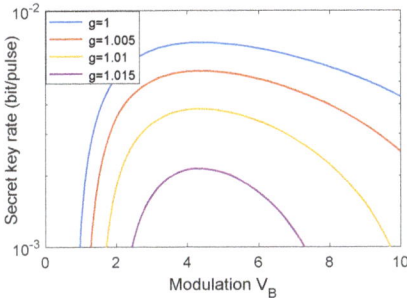

Figure A1. SCCQ protocol based on plug-and-play configuration using homodyne detection with a practical detector, d = 50, G = 1.

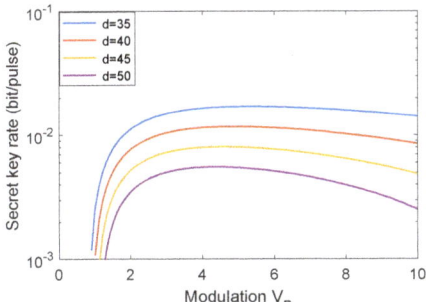

Figure A2. SCCQ protocol based on plug-and-play configuration using homodyne detection with a practical detector, g = 1.005, G = 1.

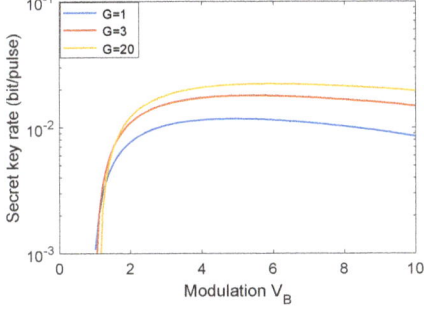

Figure A3. SCCQ protocol based on plug-and-play configuration using homodyne detection with a practical detector, g = 1.005, d = 40.

The plot of Figure A2 illustrates the relationship between secret key rate and modulation V_B under different values of distance d. We find that the peak value of secret key rate can be obtained when Bob's modulation is about 4. In other words, the optimal value of V_B in this case is 4.

Furthermore, Figure A3 shows the relationship between secret key rate and modulation V_B with different gains of the amplifiers. Here, the optimal value of Bob's modulation V_B is still about 4. That is to say that we achieve the same conclusion (optimal $V_B = 4$) as above.

In view of above analysis, we can achieve the optimal value of modulation V_B, namely, $V_B = 4$, which is deemed as a constant in our protocol.

Appendix B. Calculation of Asymptotic Secret Key Rate

Here, we calculate the asymptotic secret key rate with reverse reconciliation under the optimal collective attack, which is given by [32]

$$K_{asym} = fI(A:B) - \chi_{AE}, \tag{A1}$$

where f is the reconciliation efficiency, $I(A:B)$ represents the Shannon mutual information shared by Alice and Bob, and χ_{AE} represents the Holevo bound of the information between Eve and Alice.

The mutual information shared by Alice and Bob $I(A:B)$ is expressed as

$$I(A:B) = \frac{1}{2}\log_2 \frac{V + \chi_{tot}}{1 + \chi_{tot}}. \tag{A2}$$

The Holevo bound of the information between Eve and Alice χ_{AE} can be calculated by [26,27]

$$\chi_{AE} = \sum_{i=1}^{2} G(\frac{v_i - 1}{2}) - \sum_{i=3}^{5} G(\frac{v_i - 1}{2}), \tag{A3}$$

where $G(x) = (x+1)\log_2(x+1) - x\log_2 x$, and the symplectic eigenvalues $v_{1,2,3,4,5}$ are given by

$$v_{1,2}^2 = \frac{1}{2}[A \pm \sqrt{A^2 - 4B}], \tag{A4}$$

where

$$A = V^2(1 - 2T) + 2T + T^2(V + \chi_{line})^2, \tag{A5}$$

$$B = T^2(V\chi_{line} + 1)^2, \tag{A6}$$

$$v_{3,4}^2 = \frac{1}{2}[C \pm \sqrt{C^2 - 4D}], \tag{A7}$$

where

$$C = \frac{A\chi_{hom} + V\sqrt{B} + T(V + \chi_{line})}{T(V + \chi_{tot})}, \tag{A8}$$

$$D = \frac{\sqrt{B}(V + \sqrt{B}\chi_{hom})}{T(V + \chi_{tot})}, \tag{A9}$$

$$v_5 = 1. \tag{A10}$$

Note that in the above equations, $V = V_B + 1$, the total channel-added noise $\chi_{line} = \frac{1}{T} - 1 + \zeta_t$ and the total noise referred to the channel input $\chi_{tot} = \chi_{line} + \frac{\chi_{hom}}{T}$.

Appendix C. Secret Key Rate in the Finite-Size Scenario

For the proposed protocol, the finite-size secret key rate is expressed as [33,34]

$$K_{fini} = \frac{n}{N}[fI(A:B) - \chi_{\epsilon_{PE}}(A:E) - \Delta(n)], \tag{A11}$$

where f and $I(A:B)$ are as the same as the aforementioned definitions. Here, N stands for the total exchanged signals and n stands for the number of signals which is taken advantage of to derive QKD. The remaining signals $r = N - n$ are utilized to estimate the parameter. ϵ_{PE} represents the failure probability of parameter estimation and $\chi_{\epsilon_{PE}}(A:E)$ represents the maximum of the Holevo information compatible with the statistics except with probability ϵ_{PE}. $\Delta(n)$ is related to the security of the privacy amplification, which is given by

$$\Delta(n) = (2dimH_A + 3)\sqrt{\frac{log_2(2/\bar{\epsilon})}{n}} + \frac{2}{n}log_2(1/\epsilon_{PB}), \tag{A12}$$

where $\bar{\epsilon}$ and ϵ_{PB} stand for the smoothing parameter and the failure probability of privacy amplification, respectively. In addition, H_A represents the Hilbert space corresponding to the Alice's raw key. In our protocol, binary bits are taken advantage of to encode the raw key, thus we have $dimH_A = 2$.

In the finite-size scenario, the covariance matrix $\Gamma_{\epsilon_{PE}}$ needs to be used to calculate $\chi_{\epsilon_{PE}}(A:E)$. Besides, $\Gamma_{\epsilon_{PE}}$ minimizes the secret key rate with a probability of at least $1 - \epsilon_{PE}$. Here, r couples of correlated variables $(x_i, y_i)_{i=1...r}$ are sampled to derive the covariance matrix $\Gamma_{\epsilon_{PE}}$. Then, a normal model is used for these correlated variables, which is given by

$$y = tx + z, \tag{A13}$$

where $t = \sqrt{T}$ and z follow a centered normal distribution with variance $\psi^2 = 1 + T\zeta_t$. On the basis of Equation (A13), Alice and Bob's data can be linked. The covariance matrix $\Gamma_{\epsilon_{PE}}$ is given by

$$\Gamma_{\epsilon_{PE}} = \begin{pmatrix} (V_B + 1)I_2 & t_{min}Z\sigma_z \\ t_{min}Z\sigma_z & (t_{min}^2 V_B + \psi_{max}^2)I_2 \end{pmatrix}, \tag{A14}$$

where t_{min} and ψ_{max}^2 represent the minimum of t and maximum of ψ^2 compatible with sampled couples except with probability $\epsilon_{PE}/2$, and $Z = \sqrt{V_B^2 + 2V_B}$. Here, we denote the Maximum-likelihood estimators as $\hat{t}$ and $\hat{\psi}^2$, which can be, respectively, expressed by

$$\hat{t} = \frac{\sum_{i=1}^{r} x_i y_i}{\sum_{i=1}^{r} x_i^2} \quad \text{and} \quad \hat{\psi}^2 = \frac{1}{r}\sum_{i=1}^{r}(y_i - \hat{t}x_i)^2. \tag{A15}$$

According to this, t_{min} (the minimum of t) and ψ_{max}^2 (the maximum of ψ^2) can be derived by

$$t_{min} \approx \hat{t} - z_{\epsilon_{PE}/2}\sqrt{\frac{\hat{\psi}^2}{rV_B}},$$

$$\psi_{max}^2 \approx \hat{\psi}^2 + z_{\epsilon_{PE}/2}\frac{\sqrt{2}\hat{\psi}^2}{\sqrt{r}}, \tag{A16}$$

where $z_{\epsilon_{PE}/2}$ is such that $1 - erf(z_{\epsilon_{PE}/\sqrt{2}})/2 = \epsilon/2$, and $erf(x) = \frac{2}{\sqrt{\pi}}\int_0^x e^{-t^2}dt$ stands for error function. In order to theoretically analyze our protocol, the expected values of $\hat{t}$ and $\hat{\psi}^2$ are, respectively, given by

$$E[\hat{t}] = \sqrt{T},$$
$$E[\hat{\psi}^2] = 1 + T\zeta_t. \tag{A17}$$

Then, we can calculate t_{min} and ψ_{max}^2 as follows:

$$t_{min} \approx \sqrt{T} - z_{\epsilon_{PE}/2}\sqrt{\frac{1+T\zeta_t}{rV_B}},$$
$$\psi_{max}^2 \approx 1 + T\zeta_t + z_{\epsilon_{PE}/2}\frac{\sqrt{2}(1+T\zeta_t)}{\sqrt{r}}. \tag{A18}$$

The optimal value for the error probabilities can be taken as being

$$\bar{\epsilon} = \epsilon_{PE} = \epsilon_{PB} = 10^{-10}. \tag{A19}$$

Then, the secret key rate in the finite-size scenario can be calculated by taking advantage of the derived bounds t_{min} and ψ_{max}^2.

References

1. Gisin, N.; Ribordy, G.; Tittel, W.; Zbinden, H. Quantum cryptography. *Rev. Mod. Phys.* **2002**, *74*, 145. [CrossRef]
2. Lo, H.-K.; Curty, M.; Tamaki, K. Secure Quantum Key Distribution. *Nat. Photon.* **2014**, *8*, 595. [CrossRef]
3. Scarani, V.; Bechmann-Pasquinucci, H.; Cerf, N.J.; Dušek, M.; Lütkenhaus, N.; Peev, M. The Security of Practical Quantum Key Distribution. *Rev. Mod. Phys.* **2009**, *81*, 1301. [CrossRef]
4. Weedbrook, C.; Pirandola, S.; García-Patrón, R.; Cerf, N.J.; Ralph, T.C.; Shapiro, J.H.; Lloyd, S. Gaussian quantum information. *Rev. Mod. Phys.* **2012**, *84*, 621. [CrossRef]
5. Shor, P.W.; Preskill, J. Simple proof of security of the BB84 quantum key distribution protocol. *Phys. Rev. Lett.* **2000**, *85*, 441. [CrossRef]
6. Lo, H.K.; Chau, H.F. Unconditional security of quantum key distribution over arbitrarily long distances. *Science* **1999**, *283*, 2050. [CrossRef] [PubMed]
7. Takeda, S.; Fuwa, M.; Van, L.P.; Furusawa, A. Entanglement Swapping between Discrete and Continuous Variables. *Phys. Rev. Lett.* **2015**, *114*, 100501. [CrossRef]
8. Huang, D.; Lin, D.; Wang, C.; Liu, W.Q.; Peng, J.Y.; Fang, S.H.; Huang, P.; Zeng, G.H. Continuous-variable quantum key distribution with 1 Mbps secure key rate. *Opt. Express* **2015**, *23*, 17511. [CrossRef] [PubMed]
9. Pirandola, S.; Ottaviani, C.; Spedalieri, G.; Weedbrook, C.; Braunstein, S.L.; Lloyd, S.; Gehring, T.; Jacobsen, C.S.; Andersen, U.L. High-rate measurement-device-independent quantum cryptography. *Nat. Photon.* **2015**, *9*, 397. [CrossRef]
10. Huang, D.; Huang, P.; Li, H.; Wang, T.; Zhou, Y.; Zeng, G. Field demonstration of a continuous-variable quantum key distribution network. *Opt. Lett.* **2016**, *41*, 3511. [CrossRef]
11. Grosshans, F.; Van A.G.; Wenger, J.; Brouri, R.; Cerf, N.J.; Grangier, P. Quantum key distribution using gaussian-modulated coherent states. *Nature* **2003**, *421*, 238. [CrossRef] [PubMed]
12. Wu, X.D.; Liao, Q.; Huang, D.; Wu, X.H.; Guo, Y. Balancing four-state continuous-variable quantum key distribution with linear optics cloning machine. *Chin. Phys. B* **2017**, *26*, 110304. [CrossRef]
13. Grosshans, F.; Grangier, P. Continuous variable quantum cryptography using coherent states. *Phys. Rev. Lett.* **2002**, *88*, 057902. [CrossRef]
14. Huang, P.; Fang, J.; Zeng, G. State-discrimination attack on discretely modulated continuous-variable quantum key distribution. *Phys. Rev. A* **2014**, *89*, 042330. [CrossRef]
15. Guo, Y.; Liao, Q.; Wang, Y.J.; Huang, D.; Huang, P.; Zeng, G.H. Performance improvement of continuous-variable quantum key distribution with an entangled source in the middle via photon subtraction. *Phys. Rev. A* **2017**, *95*, 032304. [CrossRef]
16. Huang, D.; Huang, P.; Lin, D.; Zeng, G. Long-distance continuous-variable quantum key distribution by controlling excess noise. *Sci. Rep.* **2016**, *6*, 19201. [CrossRef]

17. Jouguet, P.; Kunz-Jacques, S.; Leverrier, A. Long-distance continuous-variable quantum key distribution with a Gaussian modulation. *Phys. Rev. A* **2011**, *84*, 062317. [CrossRef]
18. Milicevic, M.; Chen, F.; Zhang, L.; Gulak, P.G. Quasi-cyclic multi-edge LDPC codes for long-distance quantum cryptography. *NPJ Quantum Inf.* **2018**, *4*, 21. [CrossRef]
19. Wang, X.; Zhang, Y.; Yu, S.; Guo, H. High speed error correction for continuous-variable quantum key distribution with multi-edge type LDPC code. *Sci. Rep.* **2018**, *8*, 10543. [CrossRef] [PubMed]
20. Kikuchi, K. Fundamentals of coherent optical fiber communications. *J. Lightwave Technol.* **2016**, *34*, 157. [CrossRef]
21. Qi, B. Simultaneous classical communication and quantum key distribution using continuous variables. *Phys. Rev. A* **2016**, *94*, 042340. [CrossRef]
22. Qi, B.; Lim, C.C.W. Noise analysis of simultaneous quantum key distribution and classical communication scheme using a true local oscillator. *Phys. Rev. Appl.* **2018**, *9*, 054008. [CrossRef]
23. Huang, D.; Huang, P.; Lin, D.-K.; Wang, C.; Zeng, G.-H. High-speed continuous-variable quantum key distribution without sending a local oscillator. *Opt. Lett.* **2015**, *40*, 3695. [CrossRef] [PubMed]
24. Qi, B.; Lougovski, P.; Pooser, R.; Grice, W.; Bobrek, M. Generating the Local Oscillator Locally in Continuous-Variable Quantum Key Distribution Based on Coherent Detection. *Phys. Rev. X* **2015**, *5*, 041009. [CrossRef]
25. Soh, D.B.S.; Brif, C.; Coles, P.J.; Lütkenhaus, N.; Camacho, R.M.; Urayama, J.; Sarovar, M. Self-Referenced Continuous-Variable Quantum Key Distribution Protocol. *Phys. Rev. X* **2015**, *5*, 041010. [CrossRef]
26. Huang, D.; Huang, P.; Wang, T.; Li, H.S.; Zhou, Y.M.; Zeng, G.H. Continuous-variable quantum key distribution based on a plug-and-play dual-phase-modulated coherent-states protocol. *Phys. Rev. A* **2016**, *94*, 032305. [CrossRef]
27. Fossier, S.; Diamanti, E.; Debuisschert, T.; Tualle-Brouri, R.; Grangier, P. Improvement of continuous-variable quantum key distribution systems by using optical preamplifiers. *J. Phys. B* **2009**, *42*, 114014. [CrossRef]
28. Zhang, H.; Fang, J.; He, G. Improving the performance of the four-state continuous-variable quantum key distribution by using optical amplifiers. *Phys. Rev. A* **2012**, *86*, 022338. [CrossRef]
29. Guo, Y.; Li, R.J.; Liao, Q.; Zhou, J.; Huang, D. Performance improvement of eight-state continuous-variable quantum key distribution with an optical amplifiers. *Phys. Lett. A* **2017**, *382*, 372–381. [CrossRef]
30. Subacius, D.; Zavriyev, A.; Trifonov, A. Backscattering limitation for fiber-optic quantum key distribution systems. *Appl. Phys. Lett.* **2005**, *86*, 011103. [CrossRef]
31. Pirandola, S.; Laurenza, R.; Ottaviani, C.; Banchi, L. Fundamental limits of repeaterless quantum communications. *Nat. Commun.* **2017**, *8*, 15043. [CrossRef] [PubMed]
32. Renner, R.; Cirac, J.I. de Finetti representation theorem for infinite-dimensional quantum systems and applications to quantum cryptography. *Phys. Rev. Lett.* **2009**, *102*, 110504. [CrossRef] [PubMed]
33. Leverrier, A.; Grosshans, F.; Grangier, P. Finite-size analysis of continuous-variable quantum key distribution. *Phys. Rev. A* **2010**, *81*, 062343. [CrossRef]
34. Furrer, F.; Franz, T.; Berta, M.; Leverrier, A.; Scholz, V.B.; Tomamichel, M.; Werner, R.F. Continuous variable quantum key distribution: finite-key analysis of composable security against coherent attacks. *Phys. Rev. Lett.* **2012**, *109*, 100502. [CrossRef] [PubMed]

© 2019 by the authors. Licensee MDPI, Basel, Switzerland. This article is an open access article distributed under the terms and conditions of the Creative Commons Attribution (CC BY) license (http://creativecommons.org/licenses/by/4.0/).

Article

On the Thermodynamic Origin of Gravitational Force by Applying Spacetime Entanglement Entropy and the Unruh Effect

Shujuan Liu [1] and Hongwei Xiong [1,2,*]

[1] College of Science, Zhejiang University of Technology, Hangzhou 31023, China; sjliu@zjut.edu.cn
[2] Wilczek Quantum Center, Department of Physics and Astronomy, Shanghai Jiao Tong University, Shanghai 200240, China
* Correspondence: xionghw@zjut.edu.cn

Received: 23 January 2019; Accepted: 16 March 2019; Published: 19 March 2019

Abstract: We consider the thermodynamic origin of the gravitational force of matter by applying the spacetime entanglement entropy and the Unruh effect originating from vacuum quantum fluctuations. By analyzing both the local thermal equilibrium and quasi-static processes of a system, we may get both the magnitude and direction of Newton's gravitational force in our theoretical model. Our work shows the possibility that the elusive Unruh effect has already shown its manifestation through gravitational force.

Keywords: spacetime entanglement entropy; Unruh effect; gravitational force; thermodynamics; holographic principle

1. Introduction

In the last decade, the investigation of spacetime entanglement [1–8] has given remarkable opportunities to consider the coalescence of quantum mechanics and gravitational force, although it is still unclear how to unify quantum mechanics and general relativity. Nevertheless, the concept of quantum entanglement has been found to connect closely with some fundamental properties of spacetime, such as vacuum quantum fluctuations [9–11], the holographic principle [12–14], and black holes [15–18].

The concept of quantum entanglement has already promoted our understanding of Boltzmann entropy and statistical thermodynamics [19–21]. For a thermodynamic system we want to study, if we consider the whole system including the external environment, the thermodynamic system is highly entangled with the external environment. In this case, the usual entropy of this thermodynamic system is in fact the entanglement entropy obtained from the reduced density matrix of this thermodynamic system [22].

In the present work, we apply both the concepts of entanglement entropy and relevant thermodynamics to consider the fundamental property of spacetime. In particular, the Unruh effect for an accelerating particle is used to consider the thermodynamic origin of gravitational force. In addition, we use a quasi-static process to consider theoretically the direction of gravitational force, which has potential application for further studies of the gravitational force for dark energy [23,24], black holes, and so on.

The paper is organized as follows. In Section 2, we give a brief introduction to the Unruh effect for the Minkowski spacetime and curved spacetime. In particular, we discuss the Unruh temperature for gravitational radiation. In Section 3, we give the finite spacetime temperature distribution of matter from the consideration of spacetime entanglement entropy and statistical thermodynamics. In

Section 4, based on the consideration of the local thermal equilibrium and a quasi-static process of a system, we give an interpretation to Newton's gravitational force and in particular the attractive characteristic. In Section 5, we consider the relativistic formula of the spacetime temperature. In the last section, we give a brief summary and discussion.

2. Vacuum Quantum Fluctuations and the Unruh Effect for Minkowski Spacetime and Curved Spacetime

The Minkowski spacetime can be specified by the distance between two nearby points in spacetime, given by:

$$ds^2 = -c^2 dt^2 + dx^2 + dy^2 + dz^2. \tag{1}$$

Even for this flat spacetime without considering the spacetime curvature of general relativity, the Minkowski spacetime has some remarkable properties when both spacetime and quantum mechanics are considered.

The confluence of special relativity and quantum mechanics will lead to nontrivial vacuum quantum fluctuations [9–11]. Although we do not know the exact property of the quantum vacuum, we may assume the existence of an extremely complex and time-dependent quantum vacuum state $|\Psi_{vacuum}\rangle$ for the quantum vacuum of the Minkowski spacetime.

The usual vacuum quantum fluctuations are considered for the existence of the zero-point energy of various quantum fields. The Casimir effect [11] between two conducting metals is due to the presence of the zero-point energy of electromagnetic field. Although there are other types of zero-point energy, the conducting metals can only change the zero-point energy of the electromagnetic field in a noticeable way. Hence, the Casimir effect is about the specified vacuum quantum fluctuations due to the electromagnetic field.

Now, we turn to consider the Unruh effect in both Minkowski spacetime and curved spacetime, which originates from the vacuum quantum fluctuations and the coupling between matter and spacetime, a little similar to the Casimir effect.

2.1. The Unruh Effect for an Accelerating Particle in Minkowski Spacetime

The Unruh effect [17,25,26] is due to vacuum quantum fluctuations of various quantum fields. For an inertial frame of reference in the Minkowski spacetime, we consider a particle with four acceleration $a^\alpha = d^2 x^\alpha / d\tau^2$ with τ the proper time. We first consider the simplest case that the particle has a specified charge so that it has a coupling with a massless scalar Bose field $\phi(t, \mathbf{r})$. The scalar field ϕ should satisfy the following equation in Minkowski spacetime,

$$\left(-\frac{1}{c^2}\frac{\partial^2}{\partial t^2} + \nabla^2\right)\phi(t, \mathbf{r}) = 0. \tag{2}$$

The quantization of this scalar field leads to:

$$\hat{\phi}(t, \mathbf{r}) \sim \int d^3 \mathbf{k} \left(\hat{a}(\mathbf{k}) f_\mathbf{k} + \hat{a}^\dagger(\mathbf{k}) f_\mathbf{k}^*\right). \tag{3}$$

Here, $f_\mathbf{k} = e^{i\mathbf{k}\cdot\mathbf{r} - iE_\mathbf{k}t/\hbar}$. $\hat{a}(\mathbf{k})$ and $\hat{a}^\dagger(\mathbf{k})$ are the annihilation and creation operators for the mode $\mathbf{k}$, respectively. The vacuum state $|0_M\rangle$ of the Minkowski spacetime satisfies the following property:

$$\hat{a}(\mathbf{k})|0_M\rangle = 0 \tag{4}$$

for all the modes denoted by $\mathbf{k}$.

For the particle with four acceleration a^α, we should use the Rindler coordinate [27–30] to consider the expansion of the field operator $\hat{\phi}(t, \mathbf{r})$. In this case, we have:

$$\hat{\phi}(t, \mathbf{r}) \sim \int d^3\mathbf{k} \left(\hat{b}(\mathbf{k}) g_\mathbf{k} + \hat{b}^\dagger(\mathbf{k}) g_\mathbf{k}^* \right). \tag{5}$$

It is worthwhile to point out that in this case, $\hat{a}(\mathbf{k}) \neq \hat{b}(\mathbf{k})$ and $f_\mathbf{k} \neq g_\mathbf{k}$ for nonzero a^α.

For this accelerating particle, it will seem that there are excitations of the ϕ field in the Minkowski spacetime because $\langle 0_M | \hat{b}^\dagger(\mathbf{k}) \hat{b}(\mathbf{k}) | 0_M \rangle \neq 0$. It is shown by Unruh [17] that:

$$\langle 0_M | \hat{b}^\dagger(\mathbf{k}) \hat{b}(\mathbf{k}) | 0_M \rangle \sim \frac{1}{e^{E_\mathbf{k}/k_B T_U} - 1}, \tag{6}$$

with:

$$T_U = \frac{\hbar a}{2\pi c k_B} \tag{7}$$

the so-called Unruh temperature. Here, the proper acceleration a in this equation is the magnitude of the four acceleration defined by:

$$a = \sqrt{\eta_{\mu\nu} a^\mu a^\nu}. \tag{8}$$

$\eta_{\mu\nu}$ is the metric of the Minkowski spacetime. Further works have verified that there are no hidden correlations in the excitations of the ϕ field, which means that the excitations are purely thermal [28].

We should note that T_U can be only observed by this accelerating particle. Hence, only at the location of this particle, there are observable thermal excitations of the ϕ field, because only at the location of this particle, there is coupling with the ϕ field in the vacuum. It is similar to calculate $\langle 0_p | \hat{a}^\dagger(\mathbf{k}) \hat{a}(\mathbf{k}) | 0_p \rangle$ with $|0_p\rangle$ defined by $\hat{b}(\mathbf{k}) |0_p\rangle = 0$ for all modes $\mathbf{k}$. For an observer at rest in the Minkowski spacetime, this means that this observer will think that there is a temperature distribution around the accelerating particle with peak temperature given by T_U.

2.2. The Unruh Effect for Curved Spacetime

The concept of Unruh temperature had been generalized to curved spacetime. This is shown clearly in [30] by Jacobson where the Unruh temperature in curved spacetime is used to give a simple derivation of Hawking temperature. Here, we give a brief introduction of the Unruh temperature for curved spacetime.

We first consider a particle with a specified charge so that it has a coupling with the ϕ field. For a curved spacetime given by:

$$ds^2 = g_{\mu\nu} dx^\mu dx^\nu, \tag{9}$$

we may also define the four acceleration in curved spacetime for this particle. The four velocity u^α is:

$$u^\alpha = \frac{dx^\alpha}{d\tau}. \tag{10}$$

The four acceleration a^α is then:

$$a^\alpha = u^\mu D_\mu u^\alpha. \tag{11}$$

Here, D_μ is a covariant derivative operator.

The magnitude of the four acceleration is:

$$a = \sqrt{g_{\mu\nu} a^\mu a^\nu}. \tag{12}$$

It is worthwhile to point out that the proper acceleration a is an invariant quantity for any observer.

In a local inertial frame of reference, it is clear that previous analysis of the Unruh effect and Unruh temperature is valid, and hence, Equation (7) can be applied to curved spacetime by replacing a

given by Equation (8) with Equation (12). It is by the application of Equations (7) and (12) that the Hawking temperature can be derived with the Unruh effect.

2.3. The Unruh Effect for Gravitational Field $h_{\mu\nu}$

For a small deviation from the Minkowski metric, we may write:

$$g_{\mu\nu} = \eta_{\mu\nu} + h_{\mu\nu}. \tag{13}$$

To leading order, we get the Ricci curvature tensor as follows,

$$R_{\mu\nu} = -\frac{1}{2}\left(\partial^2 h_{\mu\nu} - \partial_\mu \partial_\lambda h_\nu^\lambda - \partial_\nu \partial_\lambda h_\mu^\lambda + \partial_\mu \partial_\nu h_\lambda^\lambda\right) + O\left(h^2\right). \tag{14}$$

Here, $h_\mu^\lambda = \eta_{\mu\nu} h^{\nu\lambda}$. Obviously, there is gauge freedom in $h_{\mu\nu}$. Similar to the case of electromagnetic field, we may use the gauge condition to consider further the physical significance of $h_{\mu\nu}$. Using the following harmonic gauge condition:

$$\partial_\mu h_\nu^\mu = \frac{1}{2}\partial_\nu h, \tag{15}$$

the symmetric tensor $h_{\mu\nu}$ will have six free components.

With this harmonic gauge condition, in a vacuum, the Einstein field equation simplifies to:

$$\left(-\frac{1}{c^2}\frac{\partial^2}{\partial t^2} + \nabla^2\right) h_{\mu\nu} = 0. \tag{16}$$

After the harmonic gauge condition, we may still make a "residual" gauge transformation so that the solution becomes:

$$h_{\mu\nu} = \epsilon_{\mu\nu} \sin(\eta_{\alpha\beta} k^\alpha x^\beta + \varphi), \tag{17}$$

with:

$$\epsilon_{\mu\nu} = \begin{pmatrix} 0 & 0 & 0 & 0 \\ 0 & \epsilon_+ & \epsilon_\times & 0 \\ 0 & \epsilon_\times & -\epsilon_+ & 0 \\ 0 & 0 & 0 & 0 \end{pmatrix}. \tag{18}$$

Here, ϵ_+ and $\epsilon_\times$ represent two independent degrees of polarizations of gravitational waves. Similar to the quantization of electromagnetic waves, we obtain gravitons, after we quantize gravitational waves. Of course, the above discussions are about the weak field approximation, which can be applied in the present work.

By quantizing the $h_{\mu\nu}$ field and carrying out almost identical calculations, we will get the same Unruh temperature for the gravitational radiation. Of course, these considerations can be applied to the Unruh temperature for electromagnetic field as well. In the following sections, we will use Equation (7) as the effective temperature for gravitational radiation. It is well known that energy is the "charge" of the gravitational field $h_{\mu\nu}$. Hence, for any particle with $a \neq 0$, there is always nonzero Unruh temperature for gravitational radiation.

We want to emphasize two properties of the Unruh effect as follows.

1. T_U should be regarded as a peak value of a local temperature distribution in an inertial frame of reference.
2. Besides the case of an electrically-charged particle usually considered for the Unruh effect, the particle may have other types of charges. Hence, T_U may also mean the temperature for other gauge fields, such as the gravitational field. Because the gravitational field is universal for any particle, Equation (7) can be applied to the gravitational field. In this paper, the Unruh temperature is considered mainly for the gravitational field.

The purpose of this subsection is to show that for an accelerating observer, it will think that there is excitation of gravitons with the same Unruh temperature as that of the scalar field and electromagnetic field. We will show that this result gives us the chance to have close connection between the Unruh effect, spacetime temperature, and gravitational force.

3. Finite Spacetime Temperature Distribution Due to Matter

3.1. The Spacetime Quantum Fluctuations

When the sum of all zero-point energies is considered, it is well known that the vacuum energy density ϵ_V is extremely high and even divergent as a result of a rough consideration. Usually, the finite value of ϵ_V may be assumed by setting the Planck energy and Planck length as the cutoff of the quantum spacetime [27]. It is natural that this will lead to violent quantum fluctuations of the spacetime geometry [31,32] at the microscopic scale of l_p. To distinguish the vacuum quantum fluctuations introduced in the preceding section, we call it spacetime quantum fluctuation in this paper.

To give a clear picture of spacetime quantum fluctuations, we consider f_{AB} defined by:

$$f_{AB} = \sqrt{\frac{\langle \Psi_{vacuum}| d_{AB}^2 |\Psi_{vacuum}\rangle - |\langle \Psi_{vacuum}| d_{AB} |\Psi_{vacuum}\rangle|^2}{\langle \Psi_{vacuum}| d_{AB}^2 |\Psi_{vacuum}\rangle}}. \tag{19}$$

Here, d_{AB} is an operator in an inertial frame of reference to measure the spatial distance between nearby points A and B in spacetime. It is clear that f_{AB} shows the fluctuations of spacetime geometry.

If A and B are macroscopically separated, it is expected that the fluctuation f_{AB} is negligible, while below or of the order of a microscopic distance l_p, there would be significant fluctuations in f_{AB}. At the present stage, we do not know the exact value of l_p. However, the existence of spacetime quantum fluctuations [31,32] and the stable spacetime property at macroscopic scales means that there should be a distance l_p. We will give further discussion of l_p in the next section.

3.2. Spacetime Entanglement Entropy and Spacetime Temperature

We consider a sphere of radius R. The surface of this sphere divides the whole universe into two systems S_A and S_B, i.e., the interior of the sphere S_A and the external environment S_B. Without the presence of any other matter in the Minkowski spacetime, the entanglement entropy is [33]:

$$S_{entangle} = -\text{Tr}\left[\rho_A \log \rho_A\right]. \tag{20}$$

Here, $\rho_A = \text{Tr}_B(\rho)$ is the reduced density matrix with $\rho = |\Psi_{vacuum}\rangle \langle \Psi_{vacuum}|$ the density matrix for the pure state $|\Psi_{vacuum}\rangle$ of the Minkowski spacetime. It is easy to show that $S_{entangle} = -\text{Tr}\left[\rho_B \log \rho_B\right]$ with $\rho_B = \text{Tr}_A(\rho)$.

For the situation that $R \gg l_p$, it is expected that the entanglement entropy $S_{entangle}$ depends only on the property of $|\Psi_{vacuum}\rangle$ in the region of a thin spherical shell with the width of the order of l_p. Hence, it seems reasonable to assume the following conjecture of the spacetime entanglement entropy:

$$S_{entangle} \sim \frac{A_{area}}{l_p^2}. \tag{21}$$

Here, $A_{area} = 4\pi R^2$ is the area of the sphere. This is the so-called area laws for the entanglement entropy [34–36]. We have another way to understand this relation. From Equation (21), we may also regard A_{area}/l_p^2 as the number of Planck areas on the spherical surface. We will show that l_p is the Planck length in due course. It is worthwhile to point out that at the present stage, this formula does not mean directly the holographic principle because we do not consider the possible presence of matter distribution inside the sphere yet.

Now, we consider the case that there is a classical particle with mass M inside the sphere. Of course, the coupling between this particle and spacetime will lead to a change of $|\Psi_{vacuum}\rangle$ on the spherical surface. Hence, the modified entanglement entropy for the sphere becomes:

$$S_{entangle}^M \sim \frac{A_{area}}{l_p^2} + \Delta S_M. \tag{22}$$

For the usual case that the particle M only gives a slight change to the curvature of the spacetime, it is expected that $\Delta S_M \ll S_{entangle}$. However, the presence of this particle will lead to an important effect by applying the holographic principle. The holographic principle [12–14] implies that the energy Mc^2 will show its effect on the spherical surface. Combined with the first law of thermodynamics $dU = T_M dS$, we have:

$$Mc^2 \sim k_B T_M(R) \times \frac{A_{area}}{l_p^2}. \tag{23}$$

$T_M(R)$ is the effective spacetime temperature on the spherical surface. From the above equation, we have:

$$T_M(R) \sim \frac{c^2 l_p^2}{4\pi k_B} \frac{M}{R^2}. \tag{24}$$

There is another way to understand this formula. We consider an ideal case that the mass M is distributed uniformly on a surface of a sphere with a radius a little smaller than R. Assume that the spacetime temperature of this case is the same as the case we are considering. On the spherical surface, the energy within the spatial cell of area l_p^2 is:

$$\epsilon = \frac{Mc^2}{4\pi R^2/l_p^2}. \tag{25}$$

Assume the microscopic freedom of this cell is i; we have:

$$\frac{i k_B T_M}{2} = \epsilon. \tag{26}$$

Because it is expected that i is of the order of one, we will also get T_M given by Equation (24). From the result given by Equation (24), we see that our consideration is self-consistent by assuming the ideal distribution of M on the spherical surface. In a sense, this distribution of T_M is the well-known Gaussian law. Here, we give an interpretation of the Gaussian law from the holographic principle and thermodynamics. In Section 5, we will give another method to calculate T_M.

It is worthwhile to discuss the following properties of this effective spacetime temperature.

(1) In the usual case, this effective spacetime temperature is extremely small by noticing that there is a factor l_p^2 in the above equation.

(2) This effective spacetime temperature is about the spacetime and gravitational field, rather than the electromagnetic field.

(3) Because this effective spacetime temperature originates from the entanglement entropy and the presence of M inside the sphere, its finite value does not mean that there would be various radiations spontaneously. We may notice these radiations only when we have an appropriate means to experience the entanglement entropy. This is a little similar to the observation of the Casimir effect [11]. We must have two conducting metals to show the Casimir effect through the coupling with the fluctuating electromagnetic field in the quantum vacuum.

It seems that it would be extremely challenging to observe this effective spacetime temperature. However, combined with the physical picture of the Unruh effect, we will show the possibility that the simultaneous considerations of this effective spacetime temperature and the Unruh effect just lead to gravitational force.

4. Newtonian Gravitational Force Derived by the Consideration of Local Spacetime Thermal Equilibrium

4.1. Spacetime Thermal Equilibrium

We consider another fictitious particle with mass m and assume that this particle does not have any other interaction in addition to gravitational force. The particle M establishes an effective vacuum temperature field $T_M(\mathbf{r})$ given by Equation (24). Now, we consider the case that the particle m is fixed at location $\mathbf{r}$, relative to M. Because there is no relative motion between M and m, the whole system has the chance to be in spacetime thermal equilibrium. For simplicity, we consider the case that $M \gg m$. To be in spacetime thermal equilibrium, there should be another effective temperature T_m for m so that:

$$T_M(\mathbf{r}) = T_m. \tag{27}$$

When the relative location between M and m is fixed, we know that in the local inertial frame of reference for m, m has a finite acceleration. It is clear that the Unruh temperature should be calculated in a local inertial frame of reference. Hence, omitting the high-order term for the proper acceleration a, the Unruh temperature for m is:

$$T_m(\mathbf{a}) = \frac{\hbar |\mathbf{a}|}{2\pi c k_B}. \tag{28}$$

Here, $\mathbf{a} = d^2\mathbf{r}/dt^2$. We will give the exact value of T_m in Section 5. It is clear that both T_M and T_m are about gravitons, so that this equation is universal for any particle. This is one of the motivations of the analysis of the Unruh effect for gravitons in Section 2.3.

The spacetime thermal equilibrium condition (27) leads to:

$$|\mathbf{a}| = \alpha \frac{c^3 l_p^2}{2\hbar} \frac{M}{r^2}. \tag{29}$$

The coefficient α can be absorbed in the definition of Newton's gravitational constant G. Compared with Newton's law of gravitational force, we have:

$$l_p = \left(\frac{2\hbar G}{\alpha c^3}\right)^{1/2}. \tag{30}$$

We see that with the choice of $\alpha = 2$, we get the conventional gravitational constant G if we regard l_p as the Planck length. Here, we show the possibility that l_p is more fundamental than G in a sense.

In the units with $\hbar=1$ and $c=1$, we have $G = l_p^2$. We see that G decreases with the decreasing of l_p. This is due to the fact that with the decreasing of l_p, the degree of freedom increases, and hence, the effective spacetime temperature decreases on the spherical surface. The spacetime thermal equilibrium condition means that m has smaller acceleration, and equivalently smaller G.

4.2. Quasi-Static Process to Determine the Direction of Gravitational Force

Previous studies only give the magnitude of gravitational force. Now, we turn to consider the direction of gravitational force. We consider a quasi-static process by an external force $\mathbf{F}_{ext}$ so that the system is always in quasi-thermal equilibrium. In addition, we consider the case that the particle m moves toward M in a quasi-static way. Because $T_M \sim 1/r^2$, we see that the particle m will exchange heat energy with spacetime during the quasi-static process, while the kinetic energy will not change. The first law of thermodynamics then gives:

$$dU_m = \delta Q + \delta W = 0. \tag{31}$$

Here, δQ is the heat energy absorbed from spacetime, while δW is the work by the external force on the particle m. δW is given by:

$$\delta W = \mathbf{F}_{ext} \cdot d\mathbf{r}. \tag{32}$$

Hence, for the quasi-static process of the system with $dU_m = 0$, we have:

$$\mathbf{F}_{ext} \cdot d\mathbf{r} = -\delta Q. \tag{33}$$

For simplicity, we consider the case that the particle m moves along the line connecting M and m. If their distance increases, $\delta Q < 0$, and we have:

$$\mathbf{F}_{ext} \sim \frac{\mathbf{r}}{r^3}.$$

This determines the direction of the external force to maintain the thermal equilibrium or time-independent location of the particle m. We see that this is equivalent to the fact that the gravitational force is attractive.

If we consider the case that the particle m moves toward M along the line connecting M and m, we have $\delta Q > 0$. We will still have $\mathbf{F}_{ext} \sim \mathbf{r}/r^3$, and this leads to the attractive characteristic of gravitational force as well. Combined with Equation (29), the gravitational force can then be written as:

$$\mathbf{F}_g = -\frac{GMm}{r^3}\mathbf{r}. \tag{34}$$

4.3. Free-Fall Motion

In a gravitational field, we know that the free-fall motion has no acceleration at all, based on Einstein's general relativity. In this case, the Unruh temperature is zero for the particle m, while the spacetime temperature due to M is larger than zero. Hence, during the free-fall motion, there is always a temperature difference between $T_M(\mathbf{r})$ and the Unruh temperature T_m. Because of this temperature difference, the free fall motion is not a quasi-static process. This temperature difference leads to the possibility of energy exchange between spacetime and the particle m.

Similarly to the analysis of Joule expansion in thermodynamics, for the free-fall motion from A to B, we may construct a quasi-static process from A to B by a fictitious external force, and then, at the end of this quasi-static process, the work of the external force is given to the particle m. In this case, in the non-relativistic approximation, the work by the gravitational force on the particle m during the free-fall motion is:

$$\Delta W = \phi(r_1) - \phi(r_2), \tag{35}$$

with $\phi(r) = -GMm/r$ and ΔW the work done on the particle m by the gravitational field.

5. Relativistic Formula of the Spacetime Temperature T_M of a Classical Particle

In Section 3.2, based on spacetime entanglement entropy, the holographic principle, and thermodynamics, we get the non-relativistic approximation of the spacetime temperature T_M for a classical particle with mass M. It seems that it is extremely difficult to give a method to calculate $T_M(r)$ in the frame of general relativity because we do not know the exact mechanism to unify general relativity and quantum mechanics yet. However, in this section, we will provide the method to calculate $T_M(r)$ by using the local thermal equilibrium condition $T_M(\mathbf{r}) = T_m$.

For a classical particle with mass M, the Schwarzschild metric is:

$$ds^2 = -\left(1 - \frac{2GM}{c^2 r}\right)c^2 dt^2 + \left(1 - \frac{2GM}{c^2 r}\right)^{-1} dr^2 + r^2\left(d\theta^2 + \sin^2\theta d\phi^2\right). \tag{36}$$

The four-dimensional coordinate of another particle m is:

$$x^\alpha = (t, r, \theta, \phi). \tag{37}$$

We consider the situation that the particle m has a fixed location, i.e., r, θ, and ϕ are time independent. The four velocity u^α is:

$$u^\alpha = \frac{dx^\alpha}{d\tau} = \left(\frac{1}{c}\left(1 - \frac{2GM}{r}\right)^{-1/2}, 0, 0, 0\right). \tag{38}$$

The four acceleration a^α is:

$$a^\alpha = u^\mu D_\mu u^\alpha = \left(0, \frac{MG}{r^2}, 0, 0\right). \tag{39}$$

From this result, we have:

$$a^2 = g_{\mu\nu} a^\mu a^\nu = \left(\frac{MG}{r^2}\right)^2 \left(1 - \frac{2GM}{c^2 r}\right)^{-1}. \tag{40}$$

The proper acceleration is then:

$$a = \frac{MG}{r^2}\left(1 - \frac{2GM}{c^2 r}\right)^{-1/2}. \tag{41}$$

For this particle m with fixed r, θ, and ϕ, relative to M, the Unruh temperature is then:

$$T_m(a^\alpha) = \frac{\hbar G}{2\pi c k_B} \frac{M}{r^2}\left(1 - \frac{2GM}{c^2 r}\right)^{-1/2}. \tag{42}$$

We see that the factor $\left(1 - \frac{2GM}{c^2 r}\right)^{-1/2}$ is the correction due to general relativity.

From the local thermal equilibrium condition given by Equation (27), the relativistic formula for the spacetime temperature due to particle M is then:

$$T_M(r, \theta, \phi) = \frac{l_p^2}{2\pi k_B} \frac{Mc^2}{r^2}\left(1 - \frac{2GM}{c^2 r}\right)^{-1/2}. \tag{43}$$

Here, for a comparison with Equation (24), we have used the Plank length l_p in this equation. Of course, if we regard the particle M as a point particle, this formula only holds for the situation of $r > 2GM/c^2$. Compared with Equation (24), we see that the factor $\left(1 - \frac{2GM}{c^2 r}\right)^{-1/2}$ is the correction due to curved spacetime.

Compared with the calculations of the spacetime temperature T_M in Section 3.2, in this section, we give significant improvement to calculate T_M by using Equations (27) and (42). These improvements also show that the calculations in Section 3.2 are valid in the semi-relativistic approximation, and hence, this gives the support for the concept of the entanglement entropy for spacetime and the relevant thermodynamics for spacetime based on this entanglement entropy. It also implies the validity of the holographic principle and relevant thermodynamics based on the concept of entanglement entropy, because the Unruh effect does not depend on the holographic principle.

6. Potential Application to Modified Gravity

Although the present work is not a modification to Einstein's general relativity, we may consider the potential application to modified gravity in the future. Here, we consider the potential application to two modified theories of gravity as follows.

In [37–40], the modification of the inertia originated from a reconsideration of the quantum effect in the Unruh effect was considered to give a modified gravitational law, which has potential application to modified Newtonian dynamics [41]. In particular, in [39,40], the modified inertia due to the consideration of the long-wavelength of the order of the Hubble scale in Unruh radiation is used to explain the Pioneer anomaly [42]. This means that when the long-wavelength excitation of gravitons is considered, there may be significant modification to spacetime temperature considered in the present work, because the result of the present work relies on the local thermal equilibrium condition. When the long-wave mode is addressed, there is even the condensation of gravitons, similar to Bose–Einstein condensed gases.

Another potential application would be the case of massive gravity [43]. The observations of gravitational waves have given strong confinement on the graviton mass that it should be no more than 7.7×10^{-23} eV/c^2 [44], which means that the Compton wavelength of the graviton would be at least 1.6×10^{16} m. This suggests that our theory will not give a significant modification to massive gravity for the long-wave mode below 1.6×10^{16} m. However, when the cosmology evolution is addressed, we cannot exclude the possibility of significant modification due to massive gravity. Another possibility is the modification of the massive gravity to black holes [45–50], which has seen intensive studies in the last few years. Near the horizon of a black hole, the Unruh effect has close connection with Hawking radiation, and it would be interesting to consider the gravitational radiation in the Unruh effect and the relevant spacetime temperature in this work.

7. Summary and Discussion

In summary, we consider the thermodynamic origin of Newton's gravitational force by considering simultaneously the spacetime entanglement entropy, the holographic principle, thermodynamics, and the Unruh effect for an accelerating particle. Different from previous works on the thermodynamic origin of gravitational force [2,51–53], in this work, we emphasize the quantum entanglement of spacetime. In the present paper, we do not use the assumption of the displacement entropy $\Delta S \sim d$ in [2,52], which implies a more solid basis for the thermodynamic origin of Newton's gravitational force.

Currently, the direct observation of the Unruh effect is still elusive. For an electronically-charged particle, the Unruh temperature is too small to have observable electromagnetic radiation with current techniques. Most recently, a pioneering quantum simulation of coherent Unruh radiation [54] was observed based on an ultra-cold atomic system. Of course, this observation does not show directly the original Unruh effect for spacetime. In the present work, however, we show the possibility that the original Unruh effect has already shown its manifestation through gravitational force.

The purpose of this paper is to try to propose the possibility that there exists a spacetime temperature due to the curvature of spacetime because of the existence of matter. We give the general method to calculate the spacetime temperature of a classical particle by applying the thermodynamics of spacetime and the Unruh effect. Our theory suggests that the magnitude of the four acceleration for a fixed location is of equivalent importance, compared with the scalar curvature. By analyzing both the local thermal equilibrium and quasi-static processes of a system, we may give the microscopic interpretation of the attractive characteristic of classical particles, while in general relativity, this is imposed by observation [55]. It is worthwhile to point out that even in the pioneering work about the thermodynamic origin of gravitational force in [2], there is no consideration on the direction of gravitational force. In future work, we may consider the application of the spacetime temperature to black holes, e.g., the excitation of atoms falling into a black hole [56] because of the presence of the spacetime temperature in this work. Another future work may consider the influence of the spacetime temperature on the quantum correlation of matter, which would be a complementary of the recent scenario where quantum correlations are considered theoretically to affect the gravitational field [57], by emphasizing the quantum thermodynamic characteristic of work [58].

Author Contributions: Conceptualization, H.X.; methodology, H.X.; validation, H.X., and S.L.; formal analysis, H.X. and S.L.; investigation, H.X.; writing—original draft preparation, H.X.; writing—review and editing, H.X. and S.L.

Funding: This work was supported by NSFC 11175246, 11334001.

Acknowledgments: We acknowledge the discussion with Yunuo Xiong.

Conflicts of Interest: The authors declare no conflict of interest.

References

1. Holzhey, C.; Larsen, F.; Wilczek, F. Geometric and renormalized entropy in conformal field theory. *Nucl. Phys. B* **1994**, *424*, 443–467. [CrossRef]
2. Verlinde, E. On the origin of gravity and the laws of Newton. *J. High Energy Phys.* **2011**. [CrossRef]
3. Skenderis, K.; Taylor, M. The fuzzball proposal for black holes. *Phys. Rep.* **2008**, *467*, 117–171. [CrossRef]
4. Blanco, D.D.; Casini, H.; Hung, L.Y.; Myers, R.C. Relative entropy and holography. *J. High Energy Phys.* **2013**. [CrossRef]
5. Van Raamsdonk, M. Building up spacetime with quantum entanglement. *Gen. Rel. Gravit.* **2010**, *42*, 2323–2329. [CrossRef]
6. Lashkari, N.; McDermott, M.B.; Van Raamsdonk, M. Gravitational dynamics from entanglement thermodynamics. *J. High Energy Phys.* **2014**. [CrossRef]
7. Resconi, G.; Licata, I.; Fiscaletti, D. Unification of quantum and gravity by nonclassical information entropy space. *Entropy* **2013**, *15*, 3602–3619. [CrossRef]
8. Licata, I.; Chiatti, L. The archaic universe: Big bang, cosmological term and the quantum origin of time in projective cosmology. *Int. J. Theor. Phys.* **2009**, *48*, 1003–1018. [CrossRef]
9. Lamb, W.E.; Retherford, R.C. Fine structure of the hydrogen atom by a microwave method. *Phys. Rev.* **1947**, *72*, 241–243. [CrossRef]
10. Bethe, H.A. The electromagnetic shift of energy levels. *Phys. Rev.* **1947**, *72*, 339–341. [CrossRef]
11. Lamoreaux, S.K. Demonstration of the Casimir force in the 0.6 to 6 mu m range. *Phys. Rev. Lett.* **1997**, *78*, 5–8. [CrossRef]
12. Stephens, C.R.; 't Hooft, G.; Whiting, B.F. Black-hole evaporation without information loss. *Class. Quantum Grav.* **1994**, *11*, 621–647. [CrossRef]
13. Susskind, L. The world as a hologram. *J. Math. Phys.* **1995**, *36*, 6377–6396. [CrossRef]
14. Bousso, R. The holographic principle. *Rev. Mod. Phys.* **2002**, *74*, 825–874. [CrossRef]
15. Hawking, S.W. Particle creation by black-holes. *Commun. Math. Phys.* **1975**, *43*, 199–220. [CrossRef]
16. Bekenstein, J.D. Black holes and entropy. *Phys. Rev. D* **1973**, *7*, 2333–2346. [CrossRef]
17. Unruh, W.G. Notes on black-hole evaporation. *Phys. Rev. D* **1976**, *14*, 870–892. [CrossRef]
18. Almheiri, A.; Marolf, D.; Polchinski, J.; Sully, J. Black holes: complementarity or firewalls? *J. High Energy Phys.* **2013**. [CrossRef]
19. Kosloff, R. Quantum thermodynamics: A dynamical viewpoint. *Entropy* **2013**, *15*, 2100–2128. [CrossRef]
20. Lieb, E.H.; Yngvason, J. The physics and mathematics of the second law of thermodynamics. *Phys. Rep.* **1999**, *310*, 1–96. [CrossRef]
21. Brandao, F.; Horodecki, M.; Ng, N.; Oppenheim, J.; Wehner, S. The second laws of quantum thermodynamics. *Proc. Natl. Acad. Sci. USA* **2015**, *112*, 3275–3279. [CrossRef]
22. Gemmer, J.; Michel, M.; Mahler, G. *Quantum Thermodynamics*; Springer: Berlin/Heidelberg, Germany, 2009.
23. Peebles, P.J.E.; Ratra, B. The cosmological constant and dark energy. *Rev. Mod. Phys.* **2003**, *75*, 559–606. [CrossRef]
24. Xiong, H.W. On the quantitative calculation of the cosmological constant of the quantum vacuum. *arXiv* **2018**, arXiv:1805.10440.
25. Fulling, S.A. Nonuniqueness of canonical field quantization in Riemannian space-time. *Phys. Rev. D* **1973**, *7*, 2850–2862. [CrossRef]
26. Davies, P.C.W. Scalar particle production in Schwarzschild and Rindler metrics. *J. Phys. A* **1975**, *8*, 609–616. [CrossRef]
27. Zee, A. *Einstein Gravity in a Nutshell*; Princeton University Press: Princeton, NJ, USA, 2013.
28. Carroll, S. *Spacetime and Geometry*; Pearson: Edinburgh, UK, 2014.

29. Wald, R.M. *Quantum Field Theory in Curved Spacetime and Black Hole Thermodynamics*; University of Chicago Press: Chicago, IL, USA, 1994.
30. Jacobson, T.A. Introductory Lectures on Black Hole Thermodynamics; Lectures at University of Utrecht, 1996. Available online: http://www.physics.umd.edu/grt/taj/776b/lectures.pdf (accessed on 20 January 2019).
31. Wheeler, J.A. Geons. *Phys. Rev.* **1955**, *97*, 511–536. [CrossRef]
32. Power, E.A.; Wheeler, J.A. Thermal geons. *Rev. Mod. Phys.* **1957**, *29*, 480–495. [CrossRef]
33. Everett, H. Relative state formulation of quantum mechanics. *Rev. Mod. Phys.* **1957**, *29*, 454–462. [CrossRef]
34. Hawking, S.W.; Horowitz, G.T.; Ross, S.F. Entropy, area, and black-hole pairs. *Phys. Rev. D* **1995**, *51*, 4302–4314. [CrossRef]
35. Srednicki, M. Entropy and area. *Phys. Rev. Lett.* **1993**, *71*, 666–669. [CrossRef]
36. Eisert, J.; Cramer, M.; Plenio, M.B. Colloquium: area laws for the entanglement entropy. *Rev. Mod. Phys.* **2010**, *82*, 277–306. [CrossRef]
37. Milgrom, M. The modified dynamics as a vacuum effect. *Phys. Lett. A* **1999**, *253*, 273–279. [CrossRef]
38. Haisch, B.; Rueda, A.; Puthoff, H.E. Inertia as a zero-point-field Lorentz force. *Phys. Rev. A* **1993**, *49*, 678–694. [CrossRef]
39. McCulloch, M.E. Modelling the Pioneer anomaly as modified inertia. *Mon. Not. R. Astron.* **2007**, *376*, 338–342. [CrossRef]
40. McCulloch, M.E. Minimum accelerations from quantised inertia. *Europhys. Lett.* **2010**, *90*, 29001. [CrossRef]
41. Milgrom, M. A modification of the Newtonian dynamics as a possible alternative to the hidden mass hypothesis. *Astrophys. J.* **1983**, *270*, 365–370. [CrossRef]
42. Anderson, J.D.; Laing, P.A.; Lau, E.L.; Liu, A.S.; Nieto, M.M.; Turyshev, S.G. Indication, from Pioneer 10/11, Galileo and Ulysses Data, of an apparent anomalous, weak, long-range acceleration. *Phys. Rev. Lett.* **1998**, *81*, 2858–2861. [CrossRef]
43. Fierz, M.; Pauli, W. On relativistic wave equations for particles of arbitrary spin in an electromagnetic field. *Proc. R. Soc. Lond. A* **1939**, *173*, 211–232. [CrossRef]
44. Abbott, B.P.; Abbott, R.; Abbott, T.D.; Acernese, F.; Ackley, K.; Adams, C.; Adams, T.; Addesso, P.; Adhikari, R.X.; Adya, V.B.; et al. GW170104: Observation of a 50-Solar-Mass Binary Black Hole Coalescence at Redshift 0.2. *Phys. Rev. Lett.* **2017**, *118*, 221101. [CrossRef] [PubMed]
45. Cai, R.G.; Hu, Y.P.; Pan, Q.Y.; Zhang, Y.L. Thermodynamics of black holes in massive gravity. *Phys. Rev. D* **2015**, *91*, 024032. [CrossRef]
46. Arraut, I. On the apparent loss of predictability inside the de Rham-Gabadadze-Tolley non-linear formulation of massive gravity: The Hawking radiation effect. *Europhys. Lett.* **2015**, *109*, 10002. [CrossRef]
47. Arraut, I. Path-integral derivation of black-hole radiance inside the de-Rham-Gabadadze-Tolley formulation of massive gravity. *Eur. Phys. J. C* **2017**, *77*, 501. [CrossRef]
48. Hu, Y.P.; Pan, F.; Wu, X.M. The effects of massive graviton on the equilibrium between the black hole and radiation gas in an isolated box. *Phys. Lett. B* **2017**, *772*, 553–558. [CrossRef]
49. Capela, F.; Tinyakov, P.G. Black hole thermodynamics and massive gravity. *J. High Energy Phys.* **2011**, *4*, 1104. [CrossRef]
50. Capela, F.; Nardini, G. Hairy black holes in massive gravity: Thermodynamics and phase structure. *Phys. Rev. D* **2012**, *86*, 024030. [CrossRef]
51. Jacobson, T. Thermodynamics of spacetime-the Einstein equation of state. *Phys. Rev. Lett.* **1995**, *75*, 1260–1263. [CrossRef]
52. Xiong, H.W. Repulsive gravitational effect of a quantum wave packet and experimental scheme with superfluid helium. *Front. Phys.* **2015**, *10*, 1–9. [CrossRef]
53. Padmanabhan, T. Thermodynamical aspects of gravity: new insights. *Rep. Prog. Phys.* **2010**, *73*, 046901. [CrossRef]
54. Hu, J.Z.; Feng, L.; Zhang, Z.D.; Chin, C. Quantum simulation of coherent Hawking-Unruh radiation. *arXiv* **2018**, arXiv:1807.07504.
55. Weinberg, S. *Cosmology*; Oxford: New York, NY, USA, 2008.
56. Scully, M.O.; Fulling, S.; Lee, D.M.; Page, D.N.; Schleich, W.P.; Svidzinsky, A.A. Quantum optics approach to radiation from atoms falling into a black hole. *Proc. Natl. Acad. Sci. USA* **2018**, *115*, 8131–8136. [CrossRef]

57. Bruschi, D.E. On the weight of entanglement. *Phys. Lett. B* **2016**, *754*, 182–186. [CrossRef]
58. Bruschi, D.E. Work drives time evolution. *Ann. Phys.* **2018**, *394*, 155–161. [CrossRef]

© 2019 by the authors. Licensee MDPI, Basel, Switzerland. This article is an open access article distributed under the terms and conditions of the Creative Commons Attribution (CC BY) license (http://creativecommons.org/licenses/by/4.0/).

MDPI
St. Alban-Anlage 66
4052 Basel
Switzerland
Tel. +41 61 683 77 34
Fax +41 61 302 89 18
www.mdpi.com

Entropy Editorial Office
E-mail: entropy@mdpi.com
www.mdpi.com/journal/entropy

www.ingramcontent.com/pod-product-compliance
Lightning Source LLC
LaVergne TN
LVHW070601100526
838202LV00012B/529